T0283463

U N R O O T E D

UNROOTED

BOTANY,
MOTHERHOOD,
AND
THE FIGHT
TO SAVE
AN OLD
SCIENCE

ERIN ZIMMERMAN

MELVILLE HOUSE
BROOKLYN • LONDON

UNROOTED:
BOTANY, MOTHERHOOD, AND
THE FIGHT TO SAVE AN OLD SCIENCE

First published in 2024 by Melville House

First Melville House Printing: March 2024

Melville House Publishing
46 John Street
Brooklyn, NY 11201
and
Melville House UK
Suite 2000
16/18 Woodford Road
London E7 0HA

mhpbooks.com
@melvillehouse

Chapter Four originally appeared in slightly different form in *Narratively* magazine.

ISBN: 978-1-68589-070-4
ISBN: 978-1-68589-071-1(eBook)

Library of Congress Control Number: 2023950958

Designed by Beste M. Doğan

Printed in the United States of America

1 3 5 7 9 10 8 6 4 2
A catalog record for this book is available from the Library of Congress

For Clementine, Juniper, Fern & Darwin
Tout est encore possible.

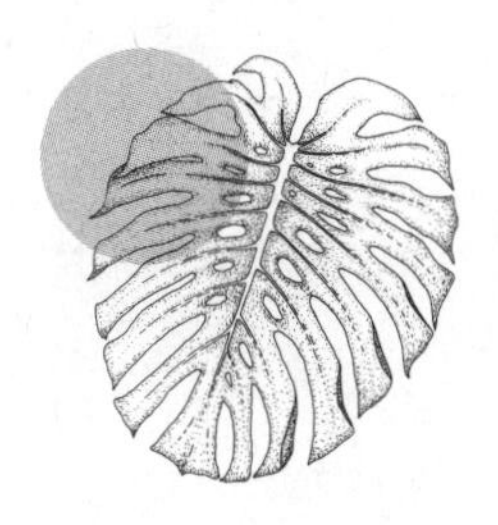

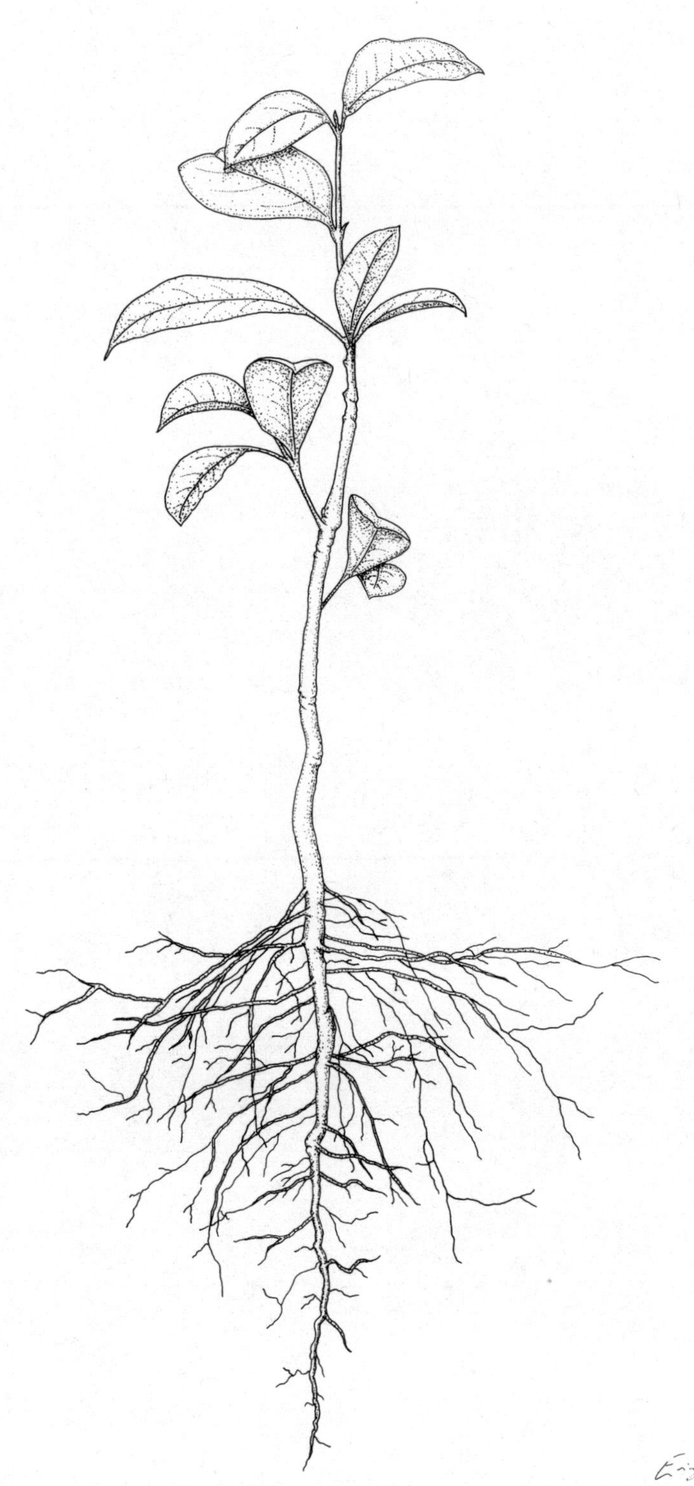

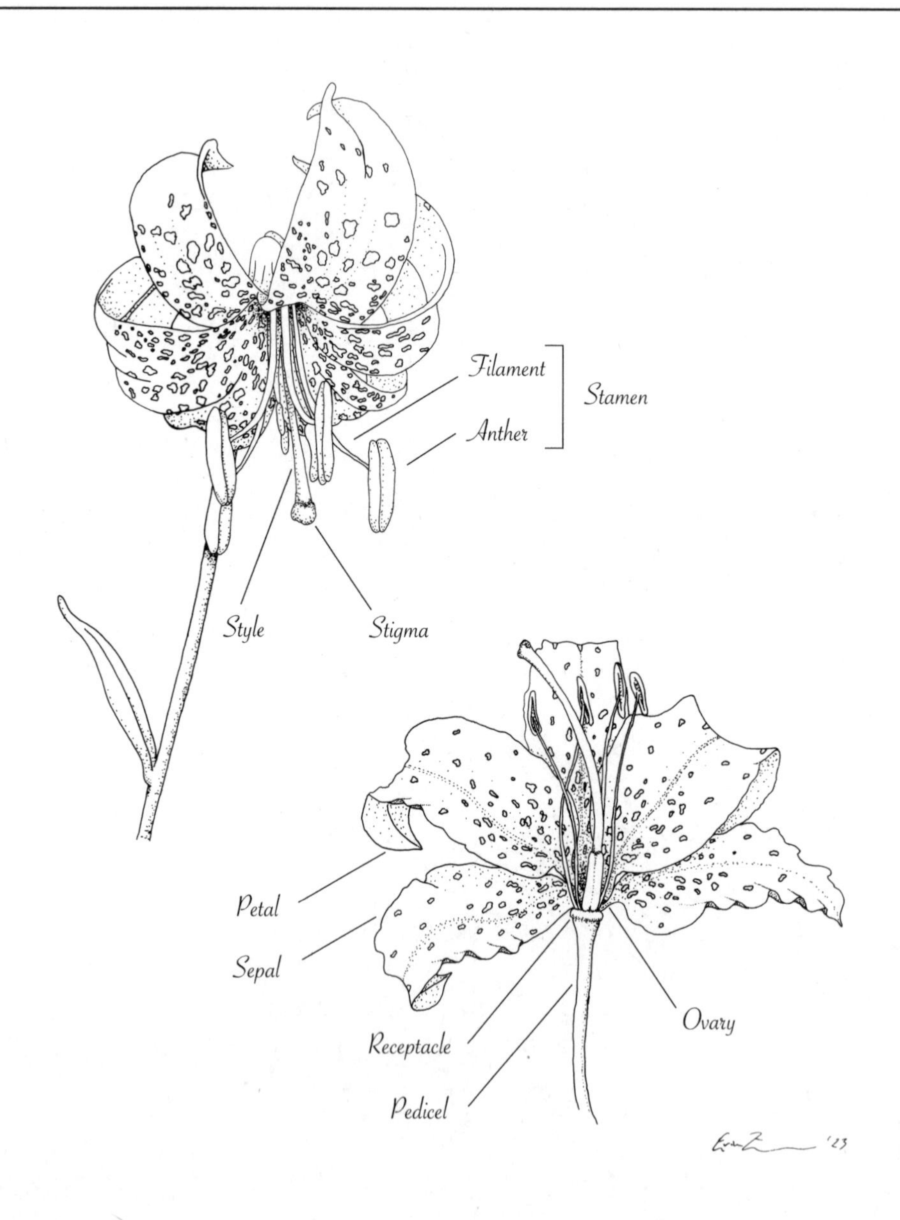
Filament
Stamen
Anther
Style
Stigma
Petal
Sepal
Receptacle
Pedicel
Ovary
'23

PROLOGUE

I sit before the end of a very long line. These few dried branches are the only evidence of a nearly unknown tree species that existed for millions upon millions of years. A quiet and unremarked goodbye that belies a much larger loss.

The specimen laid out on my workbench is a branch about a foot long, with compound leaves made up of tapered leaflets splayed out here and there and several sprigs of tiny flowers hidden among them. The whole thing is brown and brittle with age. Somehow, it even smells old. It's attached to a heavy sheet of paper, itself yellowed with time, that describes where the tree it came from was found, when, and by whom.

In the cool, airy silence of the herbarium, I hunch over to use a small hand lens I carry with me. I study the curvature of the leaves, how they are attached to the branch, and the way the veins move across them. I try to take in every detail. At length, I move to the flowers. There aren't many, and each is precious. I examine how they arise from the branch, the sort of groupings they form, before gently plucking one off with a set of tweezers and moving it to a bath of boiling water to soften the petals and other floral parts. This will allow me to dissect the flower without the brittle tissues crumbling at the touch of the tools.

The branch and flowers I'm examining are those of a tree called *Androcalymma glabrifolium*, and as far as anyone knows, it is extinct. Gone forever. I'm looking at a tiny piece of the once-living world that very few others have seen, but every flower that is taken apart to be studied is one fewer for future scientists. I try to steady my hands. This is important.

The one and only documented sighting of *Androcalymma glabrifolium*, which produced the specimen in front of me, occurred in 1936, deep in the Amazon rainforest of Brazil.[1] Because it wasn't fruiting at the time of that sighting, botanists like me have no idea what its fruit looked like. We'll probably never know. It now exists only as a few dried specimens like this, pressed on paper, tucked away in herbarium collections. I feel both privileged and sad to be one of the few scientists to have dissected one of its flowers.

It's too late for *Androcalymma*, but at least it was acknowledged and given a name before it vanished. There are countless other plant and animal species on the brink of being lost forever, and in a great many cases, we don't even know what they are yet. Science has documented about 374,000[2] of an estimated 450,000 plant species, and only about 1.2 million of an estimated 8.7 million species overall.[3]

According to the 2020 State of the World's Plants and Fungi report produced by the Royal Botanic Gardens, Kew, two plant species out of five—*40 percent of all plants*—are threatened with extinction. At least six hundred plant species are known to have been lost in modern times, and that number is likely much higher.[4]

The fossil record suggests that even during the planet's most catastrophic mass extinction events of the past, such as the global warming–induced[5] end-Permian event, which knocked out 70 percent of the world's terrestrial animal species and over 90 percent of marine invertebrates,[6] plants have fared well. They responded with adaptations to new climates and new dominant groups, but show no clear evidence of mass extinction themselves.[7] That these organisms—which couldn't be significantly impeded by what was, for animals, nearly world-ending climate change—are now massively endangered worldwide is more than a little bit frightening.

Many species are already facing extinction by the time they are first named. With habitat destruction and climate change eradicating species by the hundreds[8] each year, many will be lost before they are ever found, a phenomenon known as "dark extinction."[9]

"If something doesn't have a name, you can't conserve it," explains Damon Little, the curator of bioinformatics at the New York Botanical Garden. "You can't get a government to pay attention to it. You can't get local people to necessarily pay attention to it. You can't get a funding agency to give you money to make a seed bank or to cultivate it *in situ* in any way. Having the name is really the first step in getting any kind of conservation action to happen."

Natural history research—the collection, description, and classification of organisms—is the only way to build the foundational knowledge to understand what's being lost and what most needs protection. It also has the effect of making large, mind-numbing numbers into something real.

A hundred species of trees wiped from the face of the Earth is a statistic. Sitting silently in front of a once-living branch that exists now only as a browned and brittle carcass feels like a personal loss.

Running in parallel to the loss of both species and our capacity to build knowledge about them, a different loss of diversity is happening within scientific research. Highly trained and motivated women and other birth parents, as they have children and find that to be incompatible with their chosen career, are leaving science in droves. Like the impoverishment of genetic potential in the wake of an extinction, their departure represents a loss to research—of diverse perspectives and ways of approaching problems. As microbiologist and former National Science Foundation director Rita Colwell put it in an interview with *The Atlantic*, "Why go into battle with just half an army?"[10]

As we'll see in looking back through time to the strategies used to establish botany—and biology more broadly—as professions in the first place, the ivory tower wasn't built to accommodate women. It was explicitly built to *exclude* them, and only began to adapt as they pushed their way in. Today, academic research has become competitive to the point that, without explicit policies in place to support them, many birth parents will ultimately be forced to choose between family responsibilities and the commitment required to be sufficiently competitive in research.

It doesn't have to be this way. We know what measures are needed to stop the haemorrhage of women from the sciences. These include protected maternity leaves, designated pumping spaces, flexibility around working from home for new mothers, and childcare at conferences, so women aren't missing out on the networking and dissemination of their research that will get them hired down the road. More intangibly, a bit more understanding, particularly on the part of male researchers, of gaps and slowdowns in the publication records of female scientists with children when evaluating them for funding, hiring, or promotion would go a long way. Being the gender that bears the brunt of childrearing has concrete impacts on women's publication numbers,[11] which are a key factor in who gets chosen when one of the all-too-few tenure-track positions open up. At the level of training, male researchers train fewer women than female researchers do,[12] a gap that grows even wider the more elite those men become, effectively barring women from top labs. Put more plainly: the more successful male researchers become, the less interested they are in training women. But in a rapidly warming world charging headlong into an extinction crisis, we don't have the luxury of excluding so many of our best minds.

In 2019, nearly two thousand new plant species, mostly from South America and Asia, were named for the first time. Yet at the current rate of description and study, documenting all the unknown species projected to exist would take over a thousand years.[13]

What's more, those unknown species aren't only in remote, "unspoiled" paradises deep in the rainforest somewhere. Of the 1,942 new plant species named by scientists in 2019, over 90 were from North America. One new species found in Texas, *Eryngium arenosum*, may be useful in treating inflammation and high blood sugar.

The International Union for Conservation of Nature (IUCN) Red List of Threatened Species, usually referred to simply as the Red List, is our best and most comprehensive resource for global species extinction risk. Yet despite being the best we have, only 10 percent of plants (and 0.2 percent of fungi) have been assessed for it. Red List assessments can include various data[14] in order to make a case for a species' need for protection, but claims are primarily built around factors such as population size, generation time and life cycle, range size, changes in habitat, and specific threats such as poaching. These are data points that can be contributed by local citizen scientists for inclusion in assessments that can then be used to protect local species as needed, meaning people with no special scientific background can play an important role in fighting against species loss.

I transfer the softened flower to the dissecting microscope, where I can get a much closer look at it. This tree species survived on our planet for aeons, persevering through natural disasters, climatic shifts, and for a while, the rise of humanity, only to meet its end in the twentieth century. What can I learn from what little of it is left behind? I move through the flower one section at a time—the protective sepals, the delicate petals, the stamens and carpel that produce the pollen and fruit, respectively. How many? What shapes? What position? The sepals are five, furry, curved backward over themselves. Now the petals. Five, white, overlapping but unattached to one another. Stamens: five—this number is a common one in this group of plants—with the pollen-containing anthers bent forward and resembling a hood drawn over a head, from which comes the plant's name, *Androcalymma*: man's cowl. The carpel is single, with a long, necklike style and a pincushion stigma where the pollen lands at the tip. I note each characteristic on a sheet devoted to this species. There are well over one hundred different traits to be noted and described. Individually, these facts are unimportant, but taken as a group, they define a plant species as surely as each of your features taken together define you.

Alongside my study of its leaves and flowers, I will try and ultimately fail to extract usable DNA from the *Androcalymma* specimen to help me understand it better. But I don't know what technologies we may one day have that will allow us to learn more about this lost piece of life. In the future, we may be able to do more, so every little bit of it is precious.

The knowledge base of natural history is under threat as research funding is increasingly focussed on fast-paced, short-term experimental work over the slower-paced, longer-term observational work necessary to build and maintain it. I felt compelled to write this book because it seems to be a problem that everyone in biological research and almost no one outside of it is aware of. Like many of the extinctions quietly proceeding around the world, it just isn't something we hear about. We as citizens and stewards of this planet owe it to ourselves and our children to be aware not only of the issue but of the opportunities we have to contribute to its solution.

58
BOTANY DEPT.

CHAPTER 1

LIBRARY OF THE DEAD

Like the work of some mighty genie of Oriental fable, the brazen tube is the key that unlocks a world of wonder and beauty before invisible, which one who has once gazed upon it can never forget, and never cease to admire.

—PHILIP HENRY GOSSE, **Evenings at the microscope; or researches among the minuter forms of animal life.**

"How would you like to spend some time at Kew Gardens?" my PhD advisor, Anne, asked me in her office one afternoon. A knowing grin on her face indicated she knew I'd be thrilled. A paid trip to London is not something a grad student turns down, regardless of what they'll be doing there. But this wasn't just London, this was Kew Gardens, the most well-known and prestigious botanical garden in the world.

I stared at her wide-eyed and speechless for a moment. There are people who never leave their home institution while completing their doctorate. It wasn't something I'd known or even thought to look for when seeking out a research advisor, but Anne was someone who prioritized the work experiences of her students, allowing us to undertake fieldwork and collaborations that would develop us as scientists, lead to joint publications, and look good on our CVs. At a career stage where mentorship quality is the single biggest determinant of a student's success and happiness,[1] I was lucky to work with Dr. Anne Bruneau. A soft-spoken, petite, and vastly knowledgeable French Canadian scientist in her fifties, Anne researched the evolutionary relationships of legumes out of her lab in the Montreal Botanical Garden's research institute.

For most people, the word "legume" conjures up images of beans and peas, but it's actually quite a big and diverse family, the third largest after the grasses and the orchids. The part of the family that I was concerned with was mostly made up of tropical trees, along with a few shrubs. Most of them aren't well-known outside of the regions they grow in and have only very local common names.

The Dialiinae,* pronounced "Die-ah-*lee*-in-ee," are of interest to plant scientists because they were one of the earliest parts of the legume family to evolve.[2] Piecing together the early events of legume evolution will help us to better understand the family as a whole—what forces shaped it and which evolutionary paths weren't taken. You see, the later-evolving parts of the family, those containing beans and peas, have a very consistent and specific type of flower; the basic shape doesn't change much from one species to the next because they've hit on a winning strategy. But the Dialiinae and other early legumes have flowers that vary a lot from one genus to the next and don't look much like the classic pea flower. The group may have represented a sort of "experimental" phase in the family's evolution, when no one type of flower had yet proven successful enough to diversify into a large number of species.[3]

---

* Taxonomically, the Dialiinae have now been elevated from a subtribe to the rank of a subfamily, and are henceforth called the Dialioideae, a change that occurred very near the end of my PhD. Here, I've retained the name that was in use at the time the majority of this story took place.

My PhD research involved studying both the external structure of the Dialiinae—what's referred to as their morphology—as well as their genes to develop a picture of how they evolved. I was trying to better understand the systematics—that is, the evolutionary family tree—of their ninety or so species.

A few months later, I was on a plane to Heathrow while visions of high tea and herbarium specimens danced in my head. Situated in a fashionable corner of southwest London, the Royal Botanic Gardens, Kew, as it is formally known, was founded in 1759 as a royal pleasure garden.[4] Anne had arranged for me to spend a month as a visiting scientist and work with Dr. Gerhard Prenner, who would teach me about floral ontogeny—the study of tiny, developing flower buds. As someone obsessed with life's microscopic details, I was eager to learn more about a field of study I was sure I'd come to love.

Arriving my first day at Kew, I was whisked past both tourist lineups and locked doors like some kind of nerd VIP. Our first stop was the original 1877 wing of the herbarium.

An airy space of white paint and hardwood, Kew's original herbarium is a long room with very high ceilings and galleries along each wall, like a grand library. These galleries are lined in distinctive dark red wrought iron railings, with ornate spiral staircases in the same wrought iron at each end. It is quintessentially Victorian looking. The central space in the middle of the room holds long wooden tables for laying out and inspecting specimens, while all along the walls on ground level and in the galleries are wooden cabinets filled with stacks of vouchers going back as far as 1696.[5] Other botanists, some employees of Kew and some, like me, with Visiting Scientist IDs, moved quietly around the room, searching through the cabinets and pulling specimens out as needed. Not for the first time, science was allowing me to go places and see things that other people don't.

It's a testament to the zoo-centric nature of science education that virtually everyone is familiar with the ways in which we keep animal specimens for study by science—pinned insect collections, taxidermied mammals, and octopuses in jars of spirits come immediately to mind—but relatively few people know how plant specimens are kept and studied for science.

Our main means of keeping preserved plants for study—the herbarium specimen—goes back almost six hundred years, and has changed relatively little in that time. Plants pressed and dried in the field, once received by herbaria, are then mounted with glue or sometimes even stitched to a heavy sheet of paper with a label bearing its place of origin, collector, species (if known), and perhaps a bit about the surrounding habitat. Depending on whether that specimen has been identified to species, the voucher, as it's now called, can be filed with its genus and family or left with the other unknowns to await someone with the expertise to determine what it is.

Simple in concept, but powerful in that each specimen is a data point: This species lived in this place at this time. Many data points together can be used to test all manner of hypotheses about the causes of and solutions to environmental change, as well as to answer more traditional natural history questions of how species evolved and are related to one another, which can now be addressed with an ever more refined ability to extract and sequence very old DNA from dried plants. Institutions with herbaria arrange exchanges of material for study by researchers, creating a global lending library whose ability to answer questions is increasingly valuable as we try to understand the causes and knock-on effects of the changes we see in the world around us. Dusty old cabinets of brittle brown plants have never been so cutting edge.

On some of the wooden tables running the length of the room, specimens were laid out for study by one of the researchers. I was just passing through that day, not actually working, but I couldn't help but stop for a look at what was on display. One of the vouchers looked particularly old, with the splotchy, slightly browned paper and practiced handwritten script that's always a dead giveaway. On the sheet, half a dozen or so small, brownish-green plants with narrow, almost grasslike leaves were flattened out, curved around one another to fit. A few yellowish flowers and the long, thin seed pods of a legume species dotted the page. A label in the corner full of faded handwriting read *coast of Patagonia, 1832*, and below that, *C. Darwin*. Whoa. The faded handwriting I was looking at was Charles Darwin's! I felt like I'd just bumped into a celebrity. The naturalist we have to thank for our understanding of how evolution works, one of the most famous scientists of all time, held that plant in his hands and wrote those very words.

We don't think of Darwin as having studied plants; as Spencer Barrett, professor emeritus of ecology and evolutionary biology at the University of Toronto puts it, "If you just talk to the average layperson about Darwin, of course the first thing they're going to [mention] is survival of the fittest . . . natural selection. Then if you say, 'What did he work on?' they'll probably make the mistake of saying Darwin's finches or something, when in fact, Darwin's finches played a very minor role." History links Darwin more with the fauna of the Galapagos. But Barrett points out that Darwin spent more of his working life studying plants than any other group of organisms. His work in botany helped him better refine his own ideas about evolution by natural selection. "He was using plants all the time to test out his ideas, and in doing so, he interpreted so much of floral form and function in the context of evolution by natural selection," says Barrett. "He certainly was a botanist. There's no question."

In fact, Darwin's research, carried out at his own home, on the genus *Primula* came very close to reproducing the Austrian monk Gregor Mendel's pea plant experiments on how genetic inheritance works, but Darwin failed to recognize their significance. Mendel's findings were published during Darwin's life but received no notice for another three decades, after both their deaths. Today, they are absolutely foundational to our understanding of genetics. Darwin's work on plants brought him very close to discovering not one but *two* key theories in biology. Even the giants of science don't knock it out of the park every time.

At the bottom of the sheet was other writing, in someone else's hand. I saw the word *Typus!*, indicating that this was a type specimen—with this collection, Darwin found a species that was new to science.

It was my first time in a herbarium, and the bar was set high.

A HERBARIUM IS A TOMB where the bodies of the dead await inspection. Hallowed ground. Arranged in their resting places alongside their closest kin, as we ourselves would be, specimens will sit for centuries in their tissue-paper shrouds, unchanging. The older ones, their labels full of the spidery cursive of an earlier time, bear the names of nations already passed into the mists of history—Rhodesia, Ceylon, Gran Colombia, Zaire. Many come from forests that no longer exist, their homes having been long ago cleared for use by the dominant primate species. For others, there was no abrupt habitat loss, only a slow fading away as their ancient range became drier or no longer supported the pollinators they required to bear young. A single specimen may stand in for its entire species, gone from the living world. Or there may be many of a kind, collected over decades or centuries, that tell a story of how life for this organism changed with the climate, the air quality, or the local ecosystem. The specimens may grow smaller over time, or rarer. They may have greater heavy metal traces in their leaves as the region passed through industrialization[6] or changing leaf anatomy as the carbon dioxide in the air around them slowly increased.[7] Those introduced to a new home without competition or predation, having mounted a successful invasion, may grow larger and become more common as time goes on. These are the stories of plant life in our world—slow but dramatic tales unfolding over many years—that the herbarium provides, if we only know how to look at the clues.

The practice of keeping collections of pressed and dried plants for study dates back to sixteenth century Italy. At that time, and right up into the Enlightenment, the study of plants was largely limited to their medicinal qualities and carried out as an offshoot of the study of medicine. It wasn't until the nineteenth century that botany fully came into its own as a stand-alone science, with plants studied for their own sake outside their utility to humans.[8] Initially, herbarium specimens were bound together as books; only later did they become the separate sheets we use today. Though several cultures—including those in China, India, and the Islamic world—had strong botanical traditions dating back millennia, the practice of keeping vast collections of dried plants for study was specific to Europe,[9] where a strongly seasonal climate (plants couldn't be observed during the winter) and a need to build a knowledge base following the Middle Ages may have encouraged its development.

Kew's herbarium recently overtook that of the Muséum National d'Histoire Naturelle in Paris as the largest in the world,[10] holding a total of over 8.1 million specimens as of 2020.[11] That number includes more than three hundred thousand type specimens,[12] the precious individual plants chosen to forever serve as references when a new species is named, making the herbarium priceless in terms of its scientific value. So crucial are type specimens to botanical science that herbaria keep them in special coloured folders in part to allow them to be quickly saved in an emergency. During the Second World War, anything that could have *possibly* been a type specimen at Kew was placed into one of these special folders for quick retrieval in case the herbarium was in danger of being destroyed.[13] They are still being sorted back out today.

The gardens are also home to twenty-seven thousand species of living plants, and are heavily involved in plant conservation activities.[14] Some of the original collections are still living there. The Palm House is home to the world's oldest potted plant, an Eastern Cape giant cycad,

Encephalartos altensteinii, that dates back to 1775.[15] In all that time, the plant has only ever bloomed once, an event witnessed by Sir Joseph Banks shortly before his death in 1820.

Banks, perhaps the greatest botanist of the eighteenth century, amassed one of the largest plant collections in history through his global network of collectors.[16] Famous thanks to his time on James Cook's *Endeavour* voyage—it was Banks who recommended Australia to the Crown as a colonial penal colony—Banks unofficially took over as director at Kew following his return in the early 1770s. He began to aggressively collect living plant specimens for the gardens with the aim of making it a world class botanical garden, comparable to those in mainland Europe. Banks and his successors through the eighteenth and nineteenth centuries employed and corresponded with collectors the world over, bringing thousands of new species to Kew, including many that are now rare or extinct elsewhere.

MY DAYS IN LONDON FELT like a dream, the sort of existence that can only last for a little while because it is almost too quaint, too perfect. I had been assigned to stay in the home of an elderly local woman named Ann O'Brien, who rented out a room specifically to female visiting scientists at Kew Gardens. That may sound awfully limiting, but she had a steady stream of tenants, some of whom stayed up to a year at a time. Ann was in her early seventies and always had immaculately coifed hair, meticulously applied lipstick, and excellent posture. A proper lady, as she would say. Her house in the neighbourhood of Kew, just a few tree-lined blocks from the garden, was small and prim and tidy, the sort of house with a postage stamp yard, doilies on the end tables, and decorative plates commemorating the Queen's Silver Jubilee on the wall. Ann didn't like to travel but loved pictures of foreign places, and her kitchen wall was plastered in dozens of postcards from former tenants who sent them from all corners of the Earth.

Ann told me she'd lived in the house since the seventies, but that in recent years, the cost of living in what had become a very posh neighbourhood necessitated taking in tenants. But she didn't mind, she said, because she enjoyed the company. In the mornings before I left, we'd take our tea and toast with marmalade in the back garden together, and in the evening, we'd sit in the living room with our dinners on our laps and watch a TV show called *University Challenge*, doing our best to answer even a few of the obscure trivia questions posed to the contestants. Ann's favourite topics of conversation were the lives of the royal family and the doings of the host of another show called *Country File*, a farmer whom she blushingly described as "dishy." I adored our evenings together.

Each night in my tiny bedroom, I typed excited missives to my boyfriend, Eric. John Oliver's younger, French Canadian doppelgänger, he was small and dark haired, extroverted, and almost maniacally cheerful.

Eric and I had met while he was doing his PhD in plant biochemistry in another lab just a floor above me at the institute in Montreal, so we had been all but joined at the hip. As both a native French speaker and a more senior graduate student, he'd often made navigating life in another culture and language feel easier. Our relationship had already produced big improvements in both his somewhat rough English and my terrible French.

We'd started dating two years earlier and had very quickly moved in together. Just four months into our relationship, while I was thousands of miles away at a conference, his apartment flooded. I'd come back to find him squatting in my home. It was meant to be temporary but worked so well that he never moved out.

Our days together had slipped easily into place; we'd wake up and drive to the lab together, meet up again at lunch as we ate with our friends, and drive home together while we talked about how our work had gone. In Eric, I had someone who understood my life and the pressures I was under completely. For someone working in a field that can seem impenetrable from the outside, taking a route through education that can seem interminable, feeling understood was everything.

HERBARIA CREATED SEVERAL ADVANTAGES FOR those who used them, explains Barbara Thiers, director emerita of the William and Lynda Steere Herbarium of the New York Botanical Gardens and author of the book *Herbarium: The Quest to Preserve & Classify the World's Plants*. They allowed people to study plants even when they couldn't see them in their living state, such as in the winter, or when some of the important parts, like flowers, wouldn't be present. Plus, plants from places separated by vast distances could be easily compared. "It created a more equal way of observing things," she says. "You could observe the features of an herb just as well as you could a tree, because they're both flattened and stored in the same space."

Today, herbaria are primarily held by major study institutions like universities, museums, and botanical gardens, but once, they were the province of individuals. "Most of the early collectors built and maintained their own herbaria," says Thiers. "The earliest institutional herbaria were really nothing more than collections of bound volumes of individuals." This trend was reflective of the individualized nature of natural history research in its early forms. Until the mid- to late nineteenth century, naturalists were primarily men of means working as self-supported amateurs. In fact, in Britain at least, taking a salary for one's scientific work was looked down upon well into the 1800s as not befitting a gentleman.[17] The lack of close ties to an institution allowed these naturalists to direct their own studies as they pleased and build their collections into valuable assets that could be sold or bequeathed when they died or retired.

Thiers believes the ultimate shift to institutional herbaria may have happened in part due to Joseph Banks's vast personal collection of over twenty-three thousand plant species, which he had maintained in his home, becoming the property of the British Museum in London after his death in 1820. It formed the kernel of the museum's growing herbarium,[18] though Banks's collections are maintained separately to this day. "Actually, there are many vestiges of separate collections," she told me. "That kind of individual aspect and separation of things by individuals isn't entirely gone from our field."

The use of dried specimens to observe and describe the world's plant species fits well with our mental picture of science in the eighteenth and nineteenth centuries, when so much was still to be discovered. In the present day, we consider that time to have passed, the major discoveries to have been made. But in herbaria, the age of discovery is still very much ongoing.

"Every expedition that goes to remote places invariably comes back with new species," says Thiers, noting that the NYBG exploratory research staff average about one hundred new species each year, ranging from fungi to flowering plants. "But indeed, a lot of that is already discovered." By "already discovered," she means the plant has been collected at some point and is sitting in a stack in a herbarium storeroom somewhere, waiting to be recognized for what it is: a species that is new to science.

How do we decide a specimen represents a new species? When a collection is brought in, it is first compared to what has been described in the published literature and to identification keys for its general grouping. If it isn't found there, a botanist might start to suspect they have something new. At that point, it needs to be compared to other herbarium specimens that seem to be most closely related. The trick is making sure that what seems new isn't just an extreme within a natural range of variation for a known species. To ensure this isn't the case, and that the plant in question really does fall outside the range of what is already known, a number of closely related specimens need to be compared.

This can be a time-consuming process and is typically carried out by taxonomists trained in a particular group of plants. The reason so many unknown species are sitting in the back rooms of the world's herbaria is that there are fewer and fewer experts trained to do that work.

Thiers tells me that the decline of such experts over the past few decades has been dramatic. "That is one of the big problems we face—that there's less of that expertise. There are certainly people with lots and lots of talent, but in order to get tenure at a university, they probably have to do a lot more than just basic discovery work, so they don't have time. People might be able to do it, but they won't, because that is not what is going to advance their career."

Thiers cites the rise of molecular biology in the move away from regarding discovery research as cutting-edge science, though she admits that this isn't a new problem. "You very rarely will find a description of new species in [a high-profile journal such as] *Nature* or *Science*, unless it's some organism like a dinosaur or something that there aren't very many of so any new one is perceived as more impactful."

With academic jobs being as competitive as they are, researchers have to do the work that will net them high-profile publications and look good when the time comes for grant applications and tenure evaluation. In a disincentivized situation such as this, it's very easy for the type of labour herbaria require to function properly to just not get done. Of course, it's reasonable to ask: What good are herbaria today? Why do we need to collect *even more* new specimens and know what they are, when most have already been collected at some point?

Since the advent of industrialization, entire species are being lost due to human activity at an alarming and increasing rate. One of the most important services a herbarium can offer is the material needed to assess a plant species' risk for extinction, allowing at-risk species to be prioritized for conservation measures such as the establishment of protected areas, breeding and maintenance in living collections such as botanical gardens (known as *ex situ* conservation), protective laws and regulations, and public awareness campaigns.[19]

Plant collections of a given species in a region can address two important indicators of its overall health—over how large an area does it occur, and how common is it within that area? If a species occurs only in a very restricted area, its security is precarious, because a natural disaster or human activity at that site could wipe out most or all of the population. Similarly, a species could exist over a wide area, but be so rare within that area that successful reproduction isn't frequent enough to maintain the population. Using data from herbarium specimens and GPS mapping software, we can ascertain both of these metrics, giving a rough estimate of a species' risk. And if the species has been well-collected over time, comparing recent collections to historical ones can indicate a growing or shrinking population. Hypothetically, if all eighty million or so of our digitized plant collections included exact geo-reference data, which many unfortunately don't, we could use computer algorithms to calculate which of the world's known, collected plant species were in danger. Of course, any data based on herbarium specimens must be checked in the field to ensure the plant wasn't simply undercollected, though this is usually not the case, Thiers tells me.

Another way herbarium specimens are useful in assessing species' risk is through the use of phenology, the study of annual seasonal events such as flowering times. The dates on which flowering plant specimens have been collected, noted on specimen labels, can reveal trends in yearly flowering times over long spans of time—centuries, in some cases. In a warming climate, many species' flowering times have started getting earlier and earlier in the season. Sometimes, this can lead to mismatches with the pollinators those plants need to survive. Thiers references a study of a European orchid and its pollinator bees: "They could see how the phenologies of these two organisms were drifting apart, and how this was eventually going to cause problems for both.

Because the bees weren't getting the pollen and the flowers weren't getting pollinated. That's a pretty good prediction of, if not extinction, at least severe stress on the population."

Herbaria therefore represent a powerful source of data for assessing extinction risk and changes in plant populations over time, but like so many constellations of natural history research, they are saddled with an image problem: that of being obsolete. As with many of the species they contain, they themselves are threatened by indifference. While the absolute numbers of herbaria worldwide are robust, with new ones opening regularly, particularly in emerging economies, old and venerable collections housing many priceless specimens look to some funding bodies like dusty old money pits that, to put it bluntly, aren't worth the cost of maintenance. As Thiers points out, herbarium budgets rarely rise in good economic times but frequently decline in bad times.

"The big threat is from a university administration or museum administration that thinks that these collections are old-fashioned, that they're just taking up a lot of space. Herbaria are not the most expensive type of natural history collection to maintain, but they are occupying space," says Thiers. What's more, they require trained staff, which are an ongoing expense—one that universities can easily decide they no longer wish to bear. Many smaller university herbaria have been closed as a means of removing a budgetary expense and repurposing much-needed buildings. "It's a constant battle, and we can never do enough to demonstrate the relevance and the importance of these data."

In recent years, herbaria have been going above and beyond to both extend the uses of their collections and to demonstrate their relevance to a curious public. The biggest move in broadening the user base is specimen digitization—a major effort, particularly over the last decade or so, to make specimen label data and high-quality images available in online public databases.

Digitization has also led to a new means of engaging the public in citizen science through the work of transcription volunteers. This involves typing the often-handwritten information on the label of a digitized specimen into a database so it can be searched. The activity has proven surprisingly popular. "I've had thousands and thousands of volunteers. At my own institution, we have contributions on the scale of forty or fifty thousand transcriptions that are donated by volunteers who like this work," says Thiers. "They like to look at old handwriting and interpret it, and they are gaining an appreciation of natural history specimens. During the pandemic time, this has been very important for people who are stuck at home, sometimes with nothing to do—this was a way to engage their minds, and they formed their own communities." Thiers points to this social benefit as reciprocally helping museums and herbaria to justify their value. "It's always key to the success of any museum or collection to be able to demonstrate your relevance, scientifically as well as societally, and having the public engaged in your collection is very important," she says. Thiers tells me that while she's thrilled at all the digitization that's happened, she worries that the move has complicated the overall task of managing and maintaining these resources. "I fear that there's sort of a sense that, well, we digitized the collections, so really it doesn't matter much what happens to the physical collections because there's a digital representation. Digitizing didn't replace anything. What it really did was add another aspect to a collection. Now, you have two collections."

Another move that, like digitization, extends the utility of specimens while also complicating their maintenance is the creation of extended specimen networks. These are herbarium vouchers that are connected via database to everything derived from or associated with them. This could include DNA samples, photographs, field observation notes, isotope measurements, associated fungal samples, and more. These are meant to form an array of data for each collection that gives a richer and more complete picture of that organism's biology, Thiers explains. This clearer picture can be key to preserving threatened species, though the movement to create the networks is still in its infancy.

"Understanding how to save species is a lot more than just knowing what has been collected where, what species are where. You have to understand . . . what are the environmental factors that are affecting it? What biotic interactions are affecting it? For many species—a very large number of species—we have very little data about these associations," Thiers says. An extended specimen network helps answer these questions by making it easier for researchers to pull together all the information that's available about a given organism, thereby giving clues as to what might be influencing its survival.

I ask Thiers if, given all the innovations in and new uses of herbaria today, she foresees a reversal of the progressive budget cuts of recent decades.

"I would love to say yes to that. I would truly love to. I don't think it's *all* a gloomy picture. There are significant investments that are made in new infrastructure for buildings," she says. However, "it is not the trend, and what we've had to do over at least the course of my career is to continually figure out how to work smarter with less resources." She says that investments tend to be temporary allotments for things like digitization or physical infrastructure, which help but don't cover the ongoing and long-term costs where the major problem lies.

"I can't foresee anything in particular that would cause the annual operating budgets of collections around the country to necessarily go up," she says. "We can only hope that they don't go down too much."

THE MAJORITY OF MY TIME at Kew was spent not in the herbarium but in the Jodrell Laboratory. The home of Kew's scientific research program, the Jodrell sits nestled among mature trees and a small pond along the southern edge of Kew's 300 acres.[60] Entering through the building's new addition, a skylit atrium and glass-sided staircases made the place feel airy and modern, in contrast to the historical feel of most of Kew's major buildings. Still, every day as I passed along the rear of the building, fresh milk in glass bottles had been delivered for the day's tea breaks, seeming charmingly old-fashioned to a North American eye.

I was there to learn about floral ontogeny, the study of developing flowers. You can think of this as the floral version of embryology, where scientists look at developing fetuses as a way of understanding similarities and differences between species. All the parts of a flower grow from what are initially very tiny bumps on the surface of what's called the floral meristem. The meristem is an area of active cell division and growth that goes through an exacting sequence of events to produce new organs. Meristems in other parts of the plant that produce leaves and branches are indeterminate, meaning they'll keep throwing off new organs as long as the plant keeps growing. Floral meristems, on the other hand, are determinate; they produce a single flower in a way that is specific to that species of plant, from the order in which the organs are produced, to their number, to the pattern in which the initial bumps are formed. All on a surface area of a few square millimetres.

This may sound incredibly fiddly and unimportant, but knowing the precise sequence of a flower's development can reveal aspects of their evolution that are only briefly visible and have passed by the time the flower opens, the way a human fetus's vestigial tail is gone by the time they are born. So for example, you might have two four-petalled flowers in a group that normally has five petals. Because of this, they look as though they may be closely related. Are they both four-petalled because they evolved from the same four-petalled ancestor? Or is it convergence—two plants that look similar but aren't closely related? If you were to look at the flowers developing and find that one initially produced five petals but the fifth never fully developed, remaining vestigial, while the other produced five petals but then two of them fused together, you'd know it was convergence, because the way they arrived at a four-petalled flower was different. You'd never be able to tell this by looking at the mature flowers.

A real-world example of this is the genus *Cassia*,[6] which is, like my study plants, part of the legume family. Originally, *Cassia* was considered to be a single genus with three different subtypes of flowers, all of which looked quite similar and all of which had instances of certain petals being enlarged compared to the rest, a feature related to their pollination by bees. However, once the development of the three different types of flowers was studied, it became apparent that the events leading to the similar flowers were quite different, including the fact that the enlarged petals were arising from different locations in the flower bud from one group to the next—they weren't the same petals. Looking at floral development made it clear that these flowers didn't all belong in the same group. They are now considered three separate genera: *Cassia*, *Senna*, and *Chamaecrista*. That's the beauty of ontogeny. You get to see not only the finished product, but the path it took to get there.

The toughest part about studying ontogeny is that developing flower buds are *small*. Really small. To see the inner organs developing, the outer protective bracts or sepals have to be dissected off with fine forceps. These are so tiny and brittle that the merest twitch of the hand while doing this could destroy the entire bud, so an early morning coffee was off the menu, as it makes your hands less steady. Floral development research is a great way to develop the skilled hands of a surgeon without having to be burdened by all that extra income.

The whole sample is around the size of a pencil lead, so a normal microscope that uses light to illuminate its subjects won't do the trick. If you want to see the most minuscule and earliest formation of a flower, you first coat it in a very thin layer of platinum, then you fire electrons—which have a much smaller wavelength than light—at it and look at the image formed when they bounce off onto detectors. The whole thing has to be done under a vacuum, because even air can disrupt the path of the electrons, lowering the image quality. The electrons penetrate where even light cannot, returning clear, sharp images at scales of a few microns. This is called scanning electron microscopy (or SEM for short), and if you've ever seen a black-and-white close-up picture of, say, a housefly's eyes, you've seen the incredibly crisp, close images SEM can produce. Since colour comes from wavelengths of visible light, not electrons, SEM images are always black and white.

We don't think of wondrous things as happening inside small, windowless rooms packed with computers and technical equipment, but I think that for scientists, this can often be the case. Each day during my work at the Jodrell lab, Gerhard and I would settle into the nondescript little room that housed the electron microscope and see the unseeable. Some days, we looked at species that had never been studied in this way before, and we were the first people to ever look inside those buds and see how those flowers formed. There is an indescribable magic to seeing something truly new to humanity, even if it's just a few square microns of a big universe.

TEATIME IN THE JODRELL WAS the perfect distillation of British tradition and botanical politics. There was a tea maker on a rotating schedule making pots of tea for everyone twice a day. People lined up according to whether they preferred "brownish water" or a strong brew. Having fetched their "cuppa," everyone arranged themselves around a few large tables in the lunchroom roughly according to age and status. As a young visiting scientist, my rank in this hierarchy was just unfixed enough to allow me to visit different tables on different days. From my point of view as an outsider gifted a few weeks on the inside of the best botanical workplace in the world, it was fascinating to see how each group thought of their place and career in science.

The youngest group was made up of the "sandwich" students doing a placement year in the midst of their undergraduate studies. Like me, their time at Kew had been handed to them, and they likely didn't yet fully understand how hard it was to have an actual job in a place like this. They discussed school and social life, music, and current events. They didn't talk about the details of their research much because it hadn't yet come to dominate their lives, but their enthusiasm for botany was evident when they mentioned interesting plant facts they'd learned or exhibits they'd visited around the gardens.

The oldest group was the senior researchers. They seemed comfortable and settled in their daily routines. They discussed mutual colleagues, upcoming conferences, and news of the scientific world. When I sat with them, they asked who my advisor was and whether I knew certain people; name dropping seemed to be a favourite pastime at this well-connected table.

The most fascinating group to observe was the one in between. The early career researchers. These were the people who knew exactly how hard it was to get where they were, how fortunate they were to be working at Kew, and how precarious it all was. Work talk seemed to be a fixation—they discussed the technical details of their research, recent publications they'd read, which scientific journals were ranked highest. Their conversations also betrayed the unsettled nature of their lives. One woman, commenting on her long commute, explained to me that none of the younger researchers at Kew could afford to live anywhere near the gardens. Another wasn't sure how long even living in London would remain sustainable. This group represented the point at which many young scientists leave academia in search of more stable ground.

Teatime in the Jodrell was in some ways the scientific world in miniature, but this would be the only time I'd be allowed to move freely among the strata.

I wrote to Eric about the anxiety and precarity I saw among the early career researchers, but of course it came as no surprise to him. The summer before I went to Kew, he had told me he was planning to leave science. We were at a friend's rented lakeside cottage in New Hampshire, enjoying a weekend away from the pressures of the lab and the shoptalk that tended to infiltrate our days at home. Summer was just beginning to give way to autumn one morning as we sat in a cheery, sunlit bedroom and he told me that he was tired of hearing about people's job struggles in research, tired of looking ahead at years of short contracts and endless job searches before having any hope of a stable position.

"But . . . you love your actual work, don't you?" I was taken aback. Our lives in science were central to both who we were and how we'd met in the first place. The version of Eric that spent his days with a pipette in hand and his evenings reading research papers was the only version I'd ever known. Who was this guy, if not an aspiring scientist? Eric and I had worked practically side by side for as long as we'd been a couple. We had an ability to understand the intricacies of each other's research that our families and friends outside the field couldn't offer. And now he was leaving.

"I love some parts of it. But the truth is, it's lonely. I want a job where I actually get to talk to other people outside my lab," he told me in his thick Quebecois accent. He stared down at the hardwood floor, the morning sun catching the whiskers on his face that were already turning salt-and-pepper in his late twenties. My mind cast about for what biology-related job he could possibly do with his skill set that would allow him to socialize the way he wanted to. A sales rep for a scientific supply company, maybe? Science doesn't put a huge premium on being gregarious and talkative.

"I think I want to be an optometrist," he said. I stared. He could have said he wanted to be a trapeze artist and I wouldn't have been more surprised. It was what he'd wanted to do before going to university, he explained, but it was an extremely competitive program, and he hadn't had the academic record it took to get in. Now, with a bachelor's degree and most of a PhD in biochemistry under his belt, he did.

"This way, I'll get to be around different people every day and have a chance to talk." He paused. "And we might have some chance of staying together." That stung. But we both knew he was right.

Couples in which both parties are in academia often have what's called the "two-body problem," a play on a planetary mechanics problem in which celestial bodies have a complex interaction of pulls on each other, rendering their movements chaotic and difficult to predict. Academic jobs are intensely competitive, and for much of your early career (at least), you don't have a great deal of choice in where you go for those jobs. You might need to travel anywhere across North America or even overseas in pursuit of a good research position, and this state of affairs can last a decade or more. Now add in a second spinning body on its own erratic path, moving independently of the first but with a mutual gravitational pull that is often at odds with either's best career path. Couples tracing this intricate dance often spend long years apart when they might otherwise have been planning and building a future together, or even starting a family. They may *never* have permanent positions at the same universities. For all but the rare few who can tolerate long-term long-distance separation, ultimately either the couple splits or someone alters their career plans.

Eric had seen our own two-body problem coming and decided to opt out. Even as I felt abandoned by his move away from research, I was also the lucky partner of someone who had been willing to bow out while I kept pursuing my scientific dreams. He said he planned to put his nearly finished PhD on hold and apply to optometry school. He would eventually finish his thesis and defend his work to earn the PhD, but he would never move on in research.

Of course, he saw the shock he'd caused. Like me, he had been conditioned to think of leaving academia as a failure, regardless of the reasoning behind it. There's a strong, if little-discussed, culture of looking at those who leave as having given up or failed. I had been guilty of it myself when friends who once planned to do a PhD ducked out after their masters. Part of me believed that they'd left because they couldn't hack it. Eric knew this as well as I did. He began to pepper me with articles about how few jobs there were in research relative to the numbers of degrees handed out, and articles about how broken our system of higher education and research is now; I eventually asked him to stop sending them to me, because each one dug me a little deeper into self-doubt and a creeping fear I'd made a massive misstep. It became something that sat between us, unspoken but ever present, a topic we didn't touch.

WHEN MY TIME AT KEW came to a close and I turned in my ID card with a heavy heart, I was at least spared the hard landing of going straight back to Montreal. Eric joined me in London, and we toured Scotland in a rented car for two weeks. Taken in the slow, lazy days of late August, the trip had an anticipatory feel, as we both knew things were about to change for us. When we returned home, Eric would immediately begin his life in medical school, a four-year path that would keep him intensely busy. He'd have a new social group and would be passing most of his days in a different part of the city—a far cry from our days commuting and eating lunch together. I was still sad at the change that was coming, but a month immersed in plants and the people who have devoted their lives to them had given me a firmer footing in my belief that I was doing the right thing, for the right reasons.

But all that was in the future, and for a little while we could just be carefree twentysomethings tooling around in our cheap rental, taking in the vistas around the lochs, ferrying to the Hebrides, riding a steam engine, and taking in some Highland games. Between sips in a string of pubs that stretched across the isles, we mused about our future together—where we'd live, how we'd try to arrange our lives so my hypothetical lab and his hypothetical optometry practice would be near one another . . . even what being married might be like. The latter was a possibility that I'd long been ambivalent about but that had come to seem more palatable of late, though I'd always been clear that children were off the table for me.

We were in the last few years of our lives when anything still felt possible, and speculation was fun and untroubled.

At a stay in a hostel set in a gorgeous fourteenth-century manor house that looked like a castle on the edge of Loch Lomond, we were separated into men's and women's dorms for a few nights. The hostel was a family-oriented one, and busy groups of parents and children were everywhere, in and out, going on hikes and playing board games with siblings. Staying in the women's section, I watched as the mothers took on the lion's share of caring for the little ones, keeping them clean and their suitcases organized, tracking down errant socks and collecting toothbrushes, among the many other unremarked tasks of managing children.

At lunch one day in the airy but bustling dining room, I made my usual throwaway remark about how much I'd hate being tied down to a bunch of kids like that. I'd made similar statements plenty of times in our relationship, but now, as Eric was set to begin his ultimate career path, it hit him differently.

"I'll change your mind. You'll see," he said, smiling knowingly.

"Why would I change my mind?" I asked, annoyed at being challenged. "It's not like I haven't had time to think about this." I was tired of people telling me my disinterest in parenthood was a phase, and it galled me that much more coming from a boyfriend.

"I'll show you what a great dad I could be, how helpful, and you'll feel differently."

His calm self-assurance was incensing. After several years together, did he really think my steadfast objection to parenthood was simply because I thought he'd stick me with all the work? And was he really so casually arrogant as to think he could change this fundamental part of my outlook? No. This wasn't arrogance. That wasn't Eric's style. This was an abiding belief in what he saw when he thought about our future together. I let him know in no uncertain terms that I would certainly not be having children and how dare he presume to try to change that. It could have been a big fight, but Eric wasn't going to die on that hill just then. He immediately let go of the whole thing, but with a placid and knowing demeanour that said I hadn't changed his thoughts on the matter at all.

WHEN TWO INDOLENT WEEKS OF hills and haggis had passed, it was time to head back to Montreal. I'd be diving into an intense period of gathering physical data from all the herbarium specimens I'd managed to locate within my group of plants—my morphological study. My time at Kew had, for the time being, lifted the weight of constant worry about my professional future; whatever ended up happening, I loved and believed in my work. The wonder of peering into developing flowers and their evolutionary past with the electron microscope had reminded me that my reasons for doing what I was doing went beyond a career plan and into the realm of a calling. I needed to understand how this hidden part of the living world worked and admire the exacting beauty of it. Moving from ontogeny to morphology meant moving from the level of electrons to that of visible light: scalpels and traditional microscopes would be the order of the day now. I was soon to find that even at this scale, each plant is a world unto itself—one you can fall right into if you spend enough hours peering through the microscope lens.

CHAPTER 2

LEARNING TO LOOK

> The whole subject is included under the general name of Morphology. This is the most interesting department of natural history, and may be said to be its very soul.
>
> —CHARLES DARWIN,
> ***On the Origin of Species by Means of Natural Selection***

Thumping up the long, echoing, subterranean stairs of the metro with the rest of the morning off-to-work crowd, I'm spit out into the cold grey of a Montreal winter. The air is still today, and everyone's breath forms a standing cloud at the door as we pour out. The sharp crunch of refrozen slush under my heavy boots tells me it's at most -10 degrees Celsius, but the sting in my lungs when I inhale tells me it's well below that. I shuffle stiff legged up the hill to avoid slipping, sidestepping icy spots on the sidewalk where I can. Above me, the place I spend most of my waking hours—Le Jardin Botanique de Montréal. The Montreal Botanical Garden. The century-old ornate brick building in front of me houses the lab where I work, but for the moment, I've got another destination in mind. Slipping around the side and in through another few sets of doors, I'm there . . . the greenhouses. It's early, and the visitors haven't begun to arrive yet, so I have the place to myself. The weak morning light of winter is gentle—pale and almost dreamlike through the glass. One by one, I take off my hat, coat, scarf, mitts, and sweater, hugging my mass of outerwear as the warm humidity of the tropical rainforest house envelops my chilled skin. The moist air is a balm to my airways after the searing dryness of the frigid outdoors. Like the cornfields of my childhood, the greenness here is so intense it feels like it's swallowing you. But instead of the burning heat of summer, this place feels like slipping into a warm bath, body and soul. I come here often in winter. It soothes me and acts as an antidote to the sterility of the molecular lab as I push through the long work of extracting DNA from my plants.

BEING BACK AT WORK WITHOUT Eric made the place feel very different. I hadn't fully appreciated the extent to which being with him was paving my way in a place where I didn't fluently speak the language. Sure, I could follow conversations, could give simple replies in French, but I couldn't . . . be *me*. For one thing, it's nearly impossible to be funny in a language that's not your own. Humour is such a subtle thing; the ability to play with words and meanings builds over time and can't easily be picked up in a tongue with its own idiosyncratic set of connotations and associations. Yet humour is so often key to making friends and building connections in a group. I went home in the evening and looked up French Canadian slang words so I could start trying to pick them out in people's speech. Sometimes I'd try to follow a Quebecois comedian delivering a routine on YouTube. It was enough to help me understand when someone else was making a joke, but never got me as far as making my own. The institute wasn't an unfriendly place, but it was hard to give people enough reason to like me beyond mere politeness as I sat and did my smile-and-nod routine each day during the lunch break.

The *Institut de recherche en biologie végétale*, or IRBV, was essentially the botany department of the University of Montreal, transplanted onto the grounds of the Montreal Botanical Garden. The research labs that ran there specialized in everything from ecology and ethnobotany to genetics and genomics. The place seemed to be run by grad students since there were very few undergraduates stationed there and faculty were often either tucked away in their offices or off teaching on campus. Despite mostly being in our midtwenties, it seemed at times like all the *real* adults had gone and left the kids in charge of this research facility. We had our own lunchroom, study cubicles, and botanical library, and because the building was connected to the greenhouses, it was easy to pop down whenever you needed an earthy pick-me-up.

I was by this time a couple of years into my PhD research and had hit my stride. My Kew research was finished and waiting to be written up for publication, and for the past half year since returning from England, I'd been working away at obtaining DNA sequences for each of my study species so I could compare them to one another. This was a tall order, given the group of species I was working on. The samples weren't always the best; because my trees were distributed across the globe and often grew scattered through remote forests, they weren't easy to collect. As such, the leaf tissue I used was often old and had been sitting for years, if not decades.

When a sample sits around for that long, the DNA strands start to degrade and break into pieces. This makes it hard to draw enough DNA out of the tissue to get a good read, and even when read, the sequences tend to be short due to the breakages—which makes them difficult to align in the way that's required for comparing them with each other, though this is an issue we're increasingly able to circumvent with recent improvements in sequencing techniques. What's more, because many of those samples were collected as herbarium specimens and not for DNA sequencing *per se*, they had sometimes been exposed to things like heat and preservative alcohol that would cause the DNA molecules to degrade even faster. So I needed to try a lot of tricks and special kits and protocols designed for old and degraded DNA samples. In many cases, I still couldn't get a good enough sample. So I worked my way through all the species I needed to sequence, the results of my efforts accrued only very slowly. Science is often like that.

Each new round of sequencing followed basically the same steps: Take a small, dried leaf sample, freeze it to a very low temperature in liquid nitrogen, and grind it to dust. Then treat the dust in a tiny test tube with chemicals that cause the DNA molecules to go into a liquid solution. At this point, centrifuging the tube will separate the DNA in liquid from the remains of the leaf dust. A different chemical will cause the DNA to come out of the solution, and you can centrifuge that down until you have a little blob of what looks like white snot. You can then draw off the extra liquid with a pipette and dry out the DNA snot until you have a minuscule white pellet of concentrated DNA molecules. Finally, you redissolve that pellet into a bit of a solution called a buffer, which is kind of a long-term protective bath for the DNA, and *voilà*, you have a purified DNA sample, ready to have its sequence read.

This molecular work was a key part of my research and would provide much of the data upon which I'd build the evolutionary trees I was working toward. The work was fascinating on an intellectual level, and seeing the DNA sequences lined up on a computer screen like something out of *The Matrix* was fun, but the work was fiddly and lacked the contact with actual plants that kept me going when the days felt long. Fortunately, my molecular work was only half of a two-pronged approach to understanding the evolutionary history of my trees.

❧

IN THE MONTHS SINCE RETURNING from Kew, I'd arranged loans from several herbaria that held specimens from my group of plants. Carefully bundled stacks of dried plants had arrived and were awaiting my attention. I felt like a kid going for their gifts on Christmas morning . . . there were many samples of species I'd only read about, and now I'd finally get to see them with my own eyes. It was time to begin my morphological study of the Dialiinae.

Morphology is the study of the structure of an organism's physical form—in the case of a plant, things like the shape of its leaves, the way its stamens are assembled and attached to the flower, the curvature of its sepals as the flower opens. For every aspect of every different type of plant, there are entire vocabularies to describe its shape and appearance, developed over centuries of comparing one species to another.

There is such an incredible range of form in the plant kingdom, from minuscule duckweeds to towering redwoods, and from common garden flowers to underground parasites like *Hydnora* that are barely recognizable as plants. Because most people tend to see plants as a "green blur"—that is, as part of the scenery rather than as distinct individuals in the way that we see animals—we also tend to assume that their appearances are somewhat decorative or arbitrary. In truth, the curves and components of a flower rival any high-end sports car for performance-driven precision. The forces of natural selection keep changes in check: if a new variation is too far from the standard model—if, say, a flower is different enough in shape that its pollinator can't recognize or manipulate the parts—that individual will quickly fail to reproduce, and the variation dies out. This is called negative selection, and it constantly prunes unsuccessful experiments from the tree of life. Evolution, far from being a climb up a ladder toward some state of perfection, is more of an aimless walk in the dark where most of the paths lead to oblivion.

Of course, if *every* experiment in form got immediately killed off, there would be no change over time—and therefore no evolution.

Legumes developed two big tricks that led to their success as a family. One is in their flowers. The part of the legume family that beans and peas belong to, the papilionoids—named for their resemblance to butterflies—evolved a flower with two petals that press together to enclose their reproductive organs. The petals have to be prised apart in order to reach the reward inside. This means that any would-be pollinator has to be strong enough to work the petals,[1] excluding many smaller insects that might try to get the nectar without serving the plants' pollination agenda. Legumes' second trick was to pair with a bacterium that can pull nitrogen, a key plant nutrient that's very limited in the soil, from the air in exchange for a protected home in the plants' roots and a share of their energy. To enable this symbiosis, legumes had to develop a special bit of root morphology called nodules to house the bacteria. This innovation makes legumes far less dependent on soil fertility than most other plants.

Changes to a plant's physical form—sometimes even relatively minor ones—can have major implications for its evolutionary success and its future as a species. The shapes we see in the greenery around us are far from arbitrary. It is therefore possible to use those tiny, incremental changes to track the path of a given species through time. That's where I came in.

In my work, I compared morphological features from different species to help me understand how closely related those species were to one another. The ultimate product of my work with both DNA and morphology would be a family tree of sorts, showing my best hypothesis on how all my species evolved relative to one another. This type of research is called systematics. The morphological data served as a counterbalance to the DNA evidence I was amassing, which was permanently missing for many of my plants and could sometimes be misleading without another form of evidence to corroborate it. There's a trend in life science research today toward ever-greater reliance on molecular-based methods—those that depend on analyzing DNA, proteins, or other chemical components—and systematics is no different. The DNA I extracted from my plants could in theory be used on its own to construct an evolutionary tree, but those species I could never get good DNA from would be left out, and any contamination or errors could very easily go undetected. Having two lines of evidence made my results more reliable.

Each morning, I arrived at the institute and set myself up on the black-topped lab bench in our bright second-floor lab with a view of the gardens. I had a stack of herbarium vouchers to one side and a stack of score sheets to the other. Each score sheet listed every feature I needed to inspect on a specimen, and codes for what form that feature took. In front of me, a binocular microscope. Branch after branch would pass under the microscope as I inspected the vegetative—that is, nonfloral—characteristics of each plant. Several specimens for each species were necessary to ensure I saw all the natural variation and wasn't just seeing an outlier, different in some way from the norm within the species. I sometimes found that previous researchers had perhaps not studied enough specimens of a given species, because there were types of variation that showed up repeatedly but were never mentioned in any previous record.

The vegetative characteristics on my list focussed largely on those of the leaves. There are more physical features to a leaf than you might think. Leaves may be compound, as in a walnut or ash, composed of many leaflets, or they may be simple, like we see in a maple. Sometimes there's more than meets the eye. A leaf may look simple but have a subtle extra joint around the base showing that it actually evolved from a compound leaf that gradually lost all but its final leaflet over time; it appears to be a simple leaf but is really something quite different. I scored the leaves for whether their margins were smooth or toothed, for the shapes of their bases and tips, and even for the curves of their veins.

After wringing every last evolutionary clue out of the leaves and branches, it was time to have a look at the flowers. Though leaves have lots of variable features, they tend to be a bit less informative because leaves have a great deal of what scientists call "phenotypic plasticity," meaning their shape can change according to the conditions they have experienced while growing, such as higher or lower light levels and nutrition and where they are on the plant. When we study and evaluate leaf characters, we try to choose only the most consistent ones. Something like size wouldn't work because it's so dependent on where and when the plant grew. Floral characters are, as we've seen, much more consistent, because they have to be—the reproductive stakes are high.

The study of the flowers began with a look at the type of inflorescence—that is, the way the flowers are grouped together. It then moved upward through the rings of floral organs: sepals, petals, stamens, carpels. Even subtle things like the order in which the petals were folded up on themselves in bud could be useful. The stamens alone had twenty different characters to be scored, right down to the shape of the tiny holes or slits that allow the pollen to escape. One of the things that made my group, the Dialiinae, fascinating is that it had relatively few commonalities tying its various species together. And some of those were in what was absent, like the inner ring of stamens that's vanished in most species, only to reappear in a few. Or the petals that have disappeared from almost the entirety of the largest genus, *Dialium*, leaving them strangely naked looking in bloom. The earliest legumes, the Dialiinae among them, have what have been described as "awkward" flowers,[2] with numerous lost organs and unique, one-off arrangements, from a period in the evolution of the family when lots of novel flower types cropped up but none were successful enough to lead to a burst of new species with more variations on the same theme. They persist today as a record of roads glanced down, but not taken.

Beyond its use in systematics research like mine, morphology is often what helps us formulate important scientific questions about plants and can be key in understanding how species are affected by the global changes brought on by human activity. However, like other natural history–based approaches, it has an image problem that costs it dearly in funding. This is an ongoing struggle for scientists like Dr. Pamela Diggle, professor and head of the Department of Ecology and Evolutionary Biology at the University of Connecticut, who focuses on the evolution of morphological diversity in her work. "Morphology is not well supported because it's viewed as old-fashioned," Diggle says. "People have been doing morphology for over two hundred years. But I would argue that the underlying knowledge of morphology is key to doing almost any other kind of biology. I don't feel like you can be a really talented geneticist if you don't understand comparative morphology. And I don't think you can be a good ecologist if you don't understand how form contributes to function." I ask her if it suffers for not being perceived as "sexy" enough in a molecular age. "Oh, absolutely, yes. It's viewed as descriptive, whereas the flash of the techniques involved in some of these other fields and the big data . . . it's just really appealing to a lot of young scientists."

"Descriptive" is a label levelled pejoratively at most forms of natural history research, implying that important discoveries can only be found by conducting experiments and analyzing results as opposed to observing and comparing parts of the natural world, then describing what's there. But the fundamental knowledge upon which experiments are *based* has been gathered in this way, and that work isn't finished. "Ultimately, a lot of what we want to be able to do is to predict what the consequences are of what we're doing to our world. If we don't understand how organisms themselves are responding, it really hobbles our ability to make those forecasts. Our generalizations are sweeping across so much unknown. Morphology is focussed on the organism itself, and that's the centre of where all of these different spheres of research need to overlap," Diggle says.

She gives the example of research she's conducting looking into how trees in temperate regions respond to changing climates. She explains that most trees and shrubs preform their leaves and flowers the previous year, then sit dormant during the winter. Once it gets sufficiently warm out, the buds set about growing and maturing. Ecologists record when plants flower or leaf out and relate that data to temperature measurements to help them understand the functioning of the ecosystem. "But," says Diggle, "people haven't even thought to ask, 'What about all the development that occurred the year before? Are there any of these environmental effects on development?' Nobody's asked, because why would you, if you don't know that that's happening?"

To try to better understand, Diggle has been working with a colleague in Alaska to study preformation as a two-year developmental cycle, manipulating temperatures during flower bud formation to see what happens the following year. "It turns out that, like almost everything, it's really complicated," she says. In some species, very warm summers will delay flowering the following year, indicating high temperatures affect early bud development. But this isn't uniform across all species. "The take-home message is that predicting plant responses to warming temperatures is not as straightforward as one would like to think when using only the kinds of data that you can record for an entire forest.

"It's a really good example of where not thinking about plants as individual organisms with their own development and their own way of making things—you're just thinking about them as a data point of when they've flowered—you lose the ability to forecast. Your margin of error goes way up, and you lose the ability to actually *understand* how forests are working."

Morphology, the branch of science Darwin declared to be the "very soul" of natural history and its "most interesting department," is ironically in danger of dying due to its own reputation for not being interesting *enough*. Despite being well-established in her career, Diggle has had to find ways to couch her morphological work in projects where it doesn't obviously take centre stage in order to attract the funding to sustain it. "I've managed to keep myself funded by making sure that my morphology spoke to something that was viewed as more, I don't know, more *current*, addressing a question that people were interested in," she says. "I got a lot of my funding for work on phenotypic plasticity, because funding right now is hitched to global change, trying to understand morphological responses to temperature. The morphology has to be done in the context of some other kind of question."

As in many areas of natural history work, expertise in morphology is built up slowly and tends to be specific to a certain group of plants. Contrast this to techniques like DNA extraction and sequencing, which, once learned, can be applied to practically any living thing, plant or otherwise. Diggle mentions a scientific manuscript she recently peer-reviewed for publication in which the authors had misunderstood the morphology of common plants used frequently as model organisms in research—in a review meant to explain the concepts to younger researchers. She says knowledge is being lost. "If the knowledge isn't passed down, then somebody has to resurrect it, and that isn't working so well.

"I wish I was more eloquent about the centrality of morphology to everything we're asking about in terms of research now. The things that you look at in science and in our leading journals—that research can't be done if we don't continue to understand the morphology of those organisms. I'm not sure how you do ecology if all the people who know how to identify organisms are dead. It can't be done. But we're losing that expertise that's going to let us understand the future of biodiversity. And I don't know what we do about it."

IT'S BEEN SAID THAT ATTENTION is the most basic expression of love. When we love something, we take the time to notice the tiny details and nuances we might otherwise have missed. The particular way a lover tilts their head when they look at you. A beloved child's long eyelashes and crooked smile. Does it also work the other way around? Do we sometimes develop a deep affection for someone or something *because* we take the time to look, and those details start to feel comforting and familiar? A seemingly nondescript plant, given the attention to learn its unique characteristics, becomes a friend we'd recognize anywhere. Once you take the time to learn the trees and their names, a walk in the woods is never lonely or alien again. The dark forest of childhood fairy tales becomes a familiar collection of maples, oaks, and beech trees, each with their own leaf shapes and bark patterns.

My plants, the Dialiinae, had my whole attention nearly every day. They were no longer just generic plants to me but individuals I recognized on sight. Taken as a group, they felt like members of a family, each with their idiosyncrasies and commonalities that tied them to the others. I had favourites, like the asymmetrical *Dicorynia guianensis,* with its big club-like stamen that probably came from several smaller stamens fusing together into one, and *Eligmocarpus cynometroides,* with its tongue-twisting name and zigzag fruit folded up like an accordion. Sitting down to my stack of herbarium specimens and alcohol-preserved flowers every day felt like losing myself in a good book.

Scientific research topics can seem narrow to the point of absurdity, like an entire career spent on a single species, but ask any scientist, and they'll tell you that there really is a lifetime's worth of discovery there. It speaks to the complexity of our universe that even the thinnest slices can be so *expansive.* To me, sustained, close attention to a little-regarded slice of that universe felt spiritual, like time spent in quiet worship before a vast and intricate cosmos, trying to know it just a little bit better.

I hadn't gotten to spend this much time in quiet communion with plants since my time working as an undergraduate. My boss, Larry, had shown up in the lab one day with stacks and stacks of dusty microscope slide boxes, some starting to come apart at the seams, each containing fifty glass slides with bits of plant tissue mounted on them. They were quite old and had been sitting in storage for who knows how long. Decades, probably. He explained to me that they had been part of a teaching collection for students to learn plant anatomy and showed all the major plant groups from the simplest mosses, horsetails, and ferns right up to the more complex forms of flowering plants. He wanted me to sit and, with a microscope-mounted camera, take representative photos of the different cell and tissue types so that he could integrate the images into his lectures.

It was a long job because there were different tissues for each type of plant that needed to be photographed, and many slides to search through to find the best and clearest representation of that tissue. It took me more than a month of eight-hour days looking through the tube of the microscope to get it all done, but in that time, I got to watch the course of plant evolution pass before my eyes. I watched breathing holes and a waxy covering appear on the plants, enabling them to survive on dry land around the same time as the first insects evolved. I watched them develop a vascular system and woody tissue so they could grow tall and upright. I watched them form seeds to preserve their young through dry spells and in arid places at a time when four-legged animals were still new. It was remarkable to sit there and see it happen. That work was part of what solidified my commitment to always study plants. How can you not love something you've watched for four hundred million years?

Examining a sample under the microscope is a wonderful way to focus the attention because your peripheral vision is gone. The rest of the world disappears, and it's just you and the object you're observing. Microscopists have always tended to develop a great enthusiasm and affection for whatever they're studying, and I think this is why. It's not so surprising, then, that microscopes were once the focal point (pun intended) of a battle between religious and secular views of life.

Microscopy is now so commonplace that we take it for granted. Even small children can observe life beyond the realm of the naked eye with an inexpensive beginner scope. The commonness of microscopes obscures what a wonder they were to people when they first became accessible and affordable to both scientists and the public. Small, mass-produced compound microscopes became affordable to middle-class families in the mid-nineteenth century. At the same time, clubs and societies began to crop up to bring together those with a shared interest in what lay beyond the visible. Books were published that acted as guides for at-home explorations of common substances, such as pond water. Surprisingly, mass-produced microscopes were objects for leisurely amusement *before* they were common scientific tools.[3] It took time and improvement of the device before researchers could envision a use for them.

Microscopes took off as a middle-class hobby at a turning point in an ever more secular society. People were increasingly open to explanations for natural phenomena that didn't call on the actions of God. The microscope and the wonder it produced pitted two sides against each other,[4] with some popularizers of science pointing to the miniature marvels as evidence of the detail and precision in God's creation, while others pointed to them as showing the mechanistic underpinnings of life. Proponents of natural theology urged people to appreciate the beauty and underlying design of microscopic life and objects as a way of better knowing God, while scientific naturalists insisted that wonder could stand on its own in a godless universe. Both sides seemed to understand that people cared about the things they saw through that captivating tube, and that the observer's interpretation of what it meant mattered in a larger sense. What you focus on—literally, in this case—you care about.

Where does that leave us today, when learning even the basic parts of a plant and how they work has been stripped out of most kids' educations? Are they going to learn to care about the green world around them? "Plants are just a hard sell," Dr. Diggle tells me. "I think, in general, when people don't see plants as living organisms, they don't fight for them. It's too easy to clear that lot or clear that forest. If you're not interested in the natural world around you—and plants are most of that natural world—it's too easy to just disregard what we're doing to it all."

Still, she sees cause for some optimism when it comes to the power of outreach to capture people's attention, particularly through the lens of perennially popular activities like gardening. She describes an activity offered to undergraduates in her department in which students were provided a pot with soil and a seedling to plant in it and take home. "And we were just overwhelmed with interest. So I think it could come back."

MY CHILDHOOD HOME WAS A century-old, red brick farmhouse sat amid the family farms, gravel roads, and deciduous forests of southern Ontario. My memories of that time are always of open spaces full of green in every direction, the drone of insects, and the smell of the crops baking in the sun. The summers were endless.

Farm life was quiet and moved with the seasons in a million little ways that gave life a reassuring and recognizable rhythm, the absence of which would ultimately make city life untenable for me. The only other houses on our road for kilometres in either direction were my aunt and uncle, a house full of cousins, and my grandparents. For a long time, I assumed everyone lived in clusters with only family for neighbours.

My parents were both math teachers at the local high school, but in their spare time, my father was a sharecropper, helping with the business of growing beans and wheat in the fields around our home, and my mother was a rabbit farmer, raising angoras for their silky wool, shearing and spinning it into skeins that I cuddled like little pets as a small child. Beyond her sixty or so rabbits, she was an enthusiastic adopter of animals, so we seemed to have at least one of everything somewhere in the barns or pastures.

My mother dots my early childhood in fleeting glimpses only. She spent much of my very early years in and out of the hospital for cancer treatments. But her love of animals was everywhere. More than her face, I remember baby rabbits, guinea pigs, goats, raccoons, horses, and any number of rescued wild and domestic animals taken in because they needed a home. My impression of her stems from how alive the farm was in my earliest recollections. Though she died of leukaemia when I was not quite five, fading into the washed-out image of a kind smile in my mind, many of her animals would live on for years, an echo of the person who'd loved them the most.

I lived a childhood largely devoid of women. I had no sisters and, by my sixth birthday, no living grandmothers, my last having died the same year as my mother. Those years were populated by my father, my much older brother, and my Slovak grandfather, who lived next door and spoke very little English, but smiled and gestured profusely to communicate. My best friend and constant playmate was a male cousin who lived down the road, a friendship that came about mostly because we were the same age and within biking distance of one another. Whatever this unusual arrangement might have cost me in nurturing relationships or an ability to relate to other women, it made me completely at ease in groups of only men, a trait that would pay off time and time again in my years in science.

As a kid, I was a collector of nature's trinkets. Among other things, I had a modest collection of animal skulls I'd found around the farm or in the woods back beyond the fields. It was years before I realized that they were mostly remains of marauding raccoons and elderly pets subject to some long-ago, too-shallow burial in the barnyard, but I was proud of my collection all the same. Being young on a farm in the last remaining pre-internet years of the late eighties and early nineties, what author Chuck Klosterman refers to as a "present-tense existence,"[5] left a lot of childhood hours to fill and a lot of time to just look, wonder, and explore.

Beyond the farm and the fields, at the back of our property, there was a small wood, run through here and there with narrow footpaths but mostly left to grow wild. One grove held the remains of a "sugar shack" my grandparents had used to make maple syrup, though there was little left beyond a rusty cauldron even then. Every evening in the summer, deer would emerge to graze at the edges of the fields, always a quick dart away from the safety of the trees. The woods both enticed and taunted me. Every year, I'd plan a "retreat" for myself, for which I intended to set up camp among the oaks and maples and sit in solitude for a day or two, sleeping in a tent at night and trying to understand what the world felt like if you just held still and listened. My father, who had long been fascinated by meditation and Eastern philosophy, certainly did nothing to discourage this kind of behavior in his daughter. I grew up with the impression that there was wisdom and enchantment to be found in sitting quietly and observing, and a retreat in the woods felt like a sort of spiritual quest to my preteen self.

But I could never see it through. When evening came and found me alone in the fading dusk, it was always too cold, too dark, and my determination would falter. I was never quite intrepid enough to be the fearless adventurer I envisioned, even a kilometre away from my own house. This has been a theme of my life; with my aggressively casual dress style and interest in things I've found on the ground, I unwittingly project an image of being outdoorsy and adventurous, the sort of girl who likes to rough it out in the woods. The disappointing reality is that I'm not. I like being warm and dry. I sunburn easily and get terrible backaches from sleeping in a tent. I am an indoorsy person with an outdoorsy fashion sense who constantly wonders why everyone thinks I want to go camping with them. It was true then, and it's true now, much as I wish it weren't.

As you'd expect from the mostly indoorsy child of two math teachers, I was bookish. Being a kid who does really well in school can be more burden than accomplishment. I found myself under mounting pressure to do something remarkable with my life. The feeling started early, with well-meaning remarks from teachers and relatives to the effect that they were sure I was going to do something special when I grew up, and that I could be anything I wanted to be (Just choose correctly!). By the time I finished high school, anything less than excellence was treated as a letdown, and high grades were more expectation than achievement. Then I began university and had to continue to do well, so I went on to grad school, and at that point, I had to find something pretty interesting to do with my career, or all that schooling would have been a waste. It was a one-way street, and all turnoffs led to disappointing someone.

Arriving at an age when I had to choose what to study at university and, like most eighteen-year-olds, having no clear idea what I wanted to do beyond studying science, I chose physics, since it was what my father had done and it seemed like it would please all the people who kept telling me I should do something special.

I was not, as it turned out, particularly good at physics. I struggled along for three years, spending each class trying to fade into the background while worrying that I'd never measure up.

But in the evenings, there were orchids.

I bought my first one, a common *Phalaenopsis*, or moth orchid, at the grocery store after chancing on an article about them in the local newspaper. I was enthralled by how singular and ethereal the blooms were: yellow shot through with fuchsia veins and a surface that shimmered at the slightest movement. I started going to orchid shows and even private breeders' homes to buy more and stranger specimens with money I saved working night shifts at a twenty-four-hour grocery store. I collected a range, from tiny pink pleurothallids that grew on a piece of bark and had to be dipped into softened water each day, to the genus that most entranced me, the *Masdevallias*. These are orchids of the South American cloud forests. Having evolved in these high-elevation forests enshrouded by clouds, the little plants with their spidery triangular blooms need both cool temperatures and high humidity, particularly at night—a difficult combination to recreate in one's home. I went to such lengths to keep my treasures happy that, one winter, I built a cold frame in a recessed basement window of the rented house I shared, a sort of low temperature mini greenhouse, to try to give them what they needed. In summer, I put them in the refrigerator each night, crowding all my food onto a single shelf so they wouldn't be too cramped. I read orchid care books, I experimented with potting mixtures, I obsessed over their health. No amount of effort seemed too much for these fairylike blooms.

After wandering through a weekend display in the lobby of the university centre, bonsai became a second botanical fixation. Miniature trees felt enthralling in much the same way the orchids did. Once a month, I'd drive an hour and a half to spend a happy evening with the thirty or so retirees who made up the Toronto Bonsai Society, nitpicking over our tiny creations and learning wiring and pruning techniques. Soon every window of my house was crowded with portents of my botanical future.

It took a long time to dawn on me that spending my days miserable in physics and my nights happy with plants might suggest that I needed a change in direction. I'd done enough to earn a minor in physics, and if I studied all summer, I could switch to a different major and only be a year late in graduating. It was the biggest thing I'd ever quit. The day I filled out the paperwork and stopped being a future physicist in my own mind, I walked home from campus in a daze, collapsed into my battered old easy chair, and cried.

After a while, I lifted my head and looked around, puffy eyed, at the absurd number of potted plants around me, including those on my three-tiered fluorescent plant stand, full of my babies. I breathed in their scent and felt a new calm wash over me. This was going to be my future. I walked over to my desk and pulled out a bright green, spiral-bound notebook with a snap closure that I'd been saving for something special. For the next two hours, I sat and wrote everything I knew about plants, and everything I wanted to know—how they adapted to their environments, how flowers could look so different from one another, how some could have no flowers at all . . . This was going to be my life.

That fall, I enrolled in a beginner-level botany course. I learned that I wasn't the only one who was captivated by tiny details, and that plants aren't the featureless mass of green that we tend to see when we don't know how to look. I learned that orchids seem so different from other plants because they have highly specialized flowers and mostly live in trees. That their blooms can mimic the bodies of insects and their roots can pull water from air. And I learned that I could love what I was learning each day.

Within a year, I'd attached myself as a summer student to the lab whose influence would define my goals and ideals for the next decade of my life—the university's plant anatomy lab. Here, I'd learn basic microscopy and help with the group's various research projects.

The lab was housed in a sixties-era red brick building that still had an original couch in the women's washroom and would soon be condemned for asbestos, but it felt like the perfect place for making botanical discoveries by microscope. It was jointly run by Larry and Usher, two of the elder botanists of the department. Usher, who studied plant structure, was a small, quiet man with a dry sense of humour that caught you by surprise, and a suite of interesting hobbies like bonsai and model trains that he mostly didn't talk about, but that hinted at a fascinating character if you could just get to know him better. Larry, who studied symbioses between plants and fungi, was taller and louder, with a playful good nature and an easy kindness about him.

Longtime tenured professors, they were both at a point in their lives and careers when their jobs and reputations were secure. Though they still worked hard and continued to publish, enjoying their days was equally important, and they created a relaxed, collegial atmosphere in their lab.

The half dozen or so technicians and students, all men other than me, worked on their projects with the air of obsessed hobbyists, gazing through microscopes intently until the constant musical background, usually punk or new wave, was broken by someone calling us all over to see what they'd found, which might be an unusual cell type or an interesting fungal formation inside a root. Either was greeted with interest and discussion. Every morning and afternoon, we all sat down for a coffee break together to discuss our work or the issues of the day. Weekends, I worked part-time jobs in the university greenhouses taking care of wheat and marigolds that were part of various experiments. My whole life quickly turned to plants.

On Friday afternoons, we retired early to the campus pub together to relax and laugh over pitchers of beer. Dispersing for dinner, we'd meet up again later at a grungy but charming downtown bar called the Albion for camaraderie that went late into the evening and, for the younger ones, most of the night. Larry retired earlier than us, but not by much. He spent hours sharing beers and joking along with the rest of us, telling wild stories of what science was like in his younger years, any line between professor and students having been erased several beers ago.

Those nights were a blur of dancing and drinking and feeling at home, as though it would always be like this. I remember one night leaning my head against the stone wall next to a window at a table by the dance floor as the others talked around me. It was spring, just as the leaves were emerging from the maple outside, lit from behind by a streetlamp. I stared at that delicate pale green glow through a pleasant haze of gin and tonics and thought that the colour of newborn leaves was the most beautiful thing in the world.

Drunk and in love in springtime is the best part of being young.

I've never been so happy to stroll into work each day as I was when I worked for Larry and Usher. To be among people I admired, looking at plant structures all day. I couldn't know then that working in that lab was giving me a very skewed idea of what being a scientist in the competitive, often underfunded rush of the twenty-first century looked like. I didn't envision that I would never again work in a lab where the researchers running it spent the bulk of their day working alongside the students and technicians, or where everyone seemed happy and relaxed and spent time together just for the enjoyment of it. It was a place I would always be trying to find again.

My entry into the world of plant science came just as my undergraduate university, the University of Guelph, was merging its Botany and Zoology departments into an Integrative Biology department in which animal studies and molecular biology would dominate and botany would nearly vanish. Faculty who focussed on natural history were retiring without being replaced. By 2011, the University of British Columbia would house the last remaining dedicated botany department in Canada.[6] In the United States, a survey by the Chicago Botanic Garden and Botanic Gardens Conservation International found that in 1988, 72 percent of the country's top fifty most-funded universities offered advanced degree programs in botany.[7] Contrary to the overall upward trend of life science degrees, by 2010, more than half had eliminated their botany programs and most related courses, dropping degrees earned in botany at the undergraduate and graduate levels in the United States by 50 percent and 41 percent respectively. Science at the level of the whole organism and the relationships between species was no longer in vogue, despite the rising tide of extinctions gaining wider recognition in the world. The irony was and is heartbreaking.

The mood in the department was one of defiance. One of the technicians in our lab printed shirts with a fist clutching a plant labelled *Independent Republic of Botany* to show we weren't happy about being dispersed. Joining the botany department on the eve of its closure felt like walking into my birthday party just as it was ending. I belonged here, but had arrived too late. I knew it wasn't a good time to be placing my bets on this kind of science. But like in most cases of falling in love, I wasn't that interested in taking a hard look at the facts. I just wanted to study plants.

One day over coffee, I asked Larry whether it made sense to try to be a botanist, an anatomist like he was, when there seemed to be fewer and fewer opportunities for people like us. My breath caught as I waited for an answer, afraid of what he'd say. I was so young, and my trust in him so complete that I was ready to take whatever he said as gospel truth. He paused and thought about it before uttering the words that would colour the rest of my years in science.

"There's always room for the best." He shrugged and smiled at me. "You just have to be the best."

AS PART OF THE MORPHOLOGY work for my doctorate, I flew to Chicago late one autumn to spend a couple of weeks working at the Chicago Botanic Garden. While there, I'd be encamped in the cozy basement guest bedroom of one of Anne's legume-scientist collaborators, Dr. Pat Herendeen. Staying with Pat's family gave the trip a warm, friendly feel. Graduate students usually come with the convenience of being young and adaptable enough to not need stays in private accommodations like hotels and can simply be stashed in one's extra bedroom. I liked it that way and tried my best to be a pleasant thing to have holed up in the basement.

Pat was tall and lean with bright eyes, smoke-white hair, and a bushy moustache, a middle-aged scientist whose seemingly bottomless energy and enthusiasm for his work made him seem younger. He was one of the most hardworking people I met in a field of hard workers. The purpose of my trip was to learn some of the more obscure technical terms I would need to fully describe features such as my plants' pollen types, the patterning of their veins, and the shapes of their hairlike trichomes. I'd also learn a few specialized techniques for visualizing such features: epidermal peels, which let you see the puzzle-piece arrangement of cells and stomata—breathing holes—on the leaf surface, and leaf clearing, which removes the pigment so you can stain the vasculature and see even the tiniest veinlets branching through the tissue, like the capillaries in our own hands.

In what I came to think of as Botany Bootcamp, I was awoken every morning at 6:30 by a bright red, old-style mechanical alarm clock with a picture of a rooster on its round face and metal bells on top that startled the hell out of me daily with their abrupt, high-pitched intensity. After packing our lunches and grabbing a quick breakfast in the quiet, still-sleeping house, we were off to the microscopy lab at the gardens. With the exception of a quick lunch in the staff lunchroom, it was heads down over a microscope till sunset from that point on. Pat explained structures and patterns and experimental methods, and I tried to commit them all to memory. I had only a few weeks to learn everything I needed to know, and he wasn't going to have me wasting time.

The labs at Chicago Botanic Garden are set up like a sort of aquarium where passersby from the public can saunter down a wide central aisle and watch the scientists as they work in a series of windowed labs with different research focuses. In Montreal, research goes on within the grounds of the gardens but is hidden from public view, as it is at most such facilities. At the lab in Chicago, I regularly looked up from my work to see curious tourists and jostling groups of energetic schoolchildren watching me. Most children peered in briefly before something else caught their attention and they zoomed off, but the occasional straggler would linger, watching closely and trying to figure out what I was doing. These were my favourites, and I'd smile at them and try to hold up something for them to look at, or at least do my work in a way that might be more engaging to watch than as a motionless figure peering into a microscope. It was a good reminder of how lucky I was to be there, on my side of the glass.

Though feeling like one of the garden's exhibits took some getting used to, I loved that visitors might leave the gardens with a mental picture of me, a twentysomething with scarlet-dyed hair and a nose piercing, bent over a microscope when they thought of a plant scientist. Scientific research is so often hidden from view that many people have only the mental images they acquired as children when it comes to what working scientists look like, and this is more often than not an older white man in a lab coat. And of course there are many of those, but the reality is so much more diverse, as any productive ecosystem should be.

One of the facets of learning morphology that is both exciting and daunting is that it comes with its own language. Some would call it jargon, but to me, these words read like an arcane spell book. Use the right combination and you can conjure a life form that is totally unique among the Earth's species. Every aspect of a plant's structure has a set of words that describes its possible appearance. Take the overall shape of a leaf. *Ovate. Lanceolate. Cordate. Falcate. Orbicular. Cuneate.* These are just a few of many words that can describe leaf shape. How about the veins in that leaf? *Reticulate. Palmate. Pinnate. Arcuate.* Or the microscopic hairlike trichomes on its surface? *Uniseriate. Multiseriate. Pilate. Capitate. Dendritic.* A thorough account of a plant using such precise terms could run pages, but it would give you an exact description of one particular species among hundreds of thousands of possible plants.

As my head filled with these words and the shapes they described, I felt like simply having the vocabulary caused me to be aware of facets of plant forms I'd never considered before. Do we simply not notice parts of our environment that we don't have words for? The concept of linguistic relativity suggests that language and the availability of descriptive terms influences both our thoughts and worldview. For me, knowing that the little scaly, leaflike things beneath true leaves are called stipules caused me to notice stipules whenever I looked at plants. I suppose I'd seen them before, but now that I had a word, they meant something. By not teaching people a basic working vocabulary around plants, we may be leaving them unlikely to fully notice the living world. It takes attention to learn to care about something.

Even the array of new descriptive terms I was learning felt like child's play next to my attempts to decipher the top tier of scientific arcana: the Latin species description. Every new species that is officially named must be described physically as a basis for recognizing and distinguishing it from all other organisms. Jargon is increasingly frowned on today as amounting to a form of gatekeeping by experts in a field, and certainly writing in a dead language excludes laypeople (and most scientists) very effectively. But there is logic behind it. Species have traditionally been described in Latin because in a living language the meaning of words can drift over time, until a term comes to mean something entirely different. Consider for example a word like "buxom," which in a few hundred years has progressed from meaning "obedient" to describing a large-breasted person.[8] Other examples of linguistic drift include words such as "awesome," "gay," and recently, "tablet." Since species descriptions are meant to stand indefinitely, it was considered important that their message be future proof. I imagine them as the Latin of a holy rite, delivered in hushed tones and with an air of reverence over dried leaves and branches, the old magic preserving their features in perpetuity:

> Flores in cymulis paucifloris dispositi, rhachidibus evidenter quam superioribus foliis brevioribus gracilibus basi ad 1.5mm latis basi articulatis; bracteae bracteolae caducae cicatricibus prominentibus. Sepala 5 aestivatione valde imbricata inaequalia oblonga vel ovato-oblonga circ. 3 mm longa 1.8-2 mm lata apice obtusa extus pubescentia ciliis mediis saepe densius dispositis intus glabra carnosa marginibus sparse ciliolatis; petala 5 subaequalia aestivatione valde imbricata oblonga 4-6.5 mm longa 1.7-2 mm lata apice obtusa basi unguiculata, unguibus

> ad 0.5 mm longis, extus pubescentia intus glabra et evidenter carinata; stamina 4 filamentis subcrassis saepe subclavatis rectis, 2-3 mm longis circ. 0.25 mm in medio latis glabris mox deciduis; antherae ovato-subrotundae circ. 0.6 mm longae vix apice biporosae subrectae vel deflexae subversatiles, connectivo circ. 0.2 mm longo paullum thecis obscurato, 4-locellatae at 2 loculis; ovarium vix stipitatum vel subsessile complanatum oblongum circ. 2 mm longum leviter omnino pubescens 2-3-ovulatum stylo brevissimo. Fructus non visi.[9]

These are the words that evoke the flowers of *Androcalymma glabrifolium*, late of the Amazon rainforest, now presumed extinct, along with their cowl-headed anthers and unknown fruit, yet preserved forever in a language that itself died out long ago.

AS SUPPERTIME ROLLED AROUND AND the laboratories emptied out, Pat and I would call it a day, driving back to his house while chatting about what I'd learned and what the next day's priority should be. My eyes burned and my head throbbed from so many hours of focussing down the barrel of a microscope, but I was learning so much I didn't care. This was how I was going to learn to be "the best," as Larry had put it. It was rare for a graduate student to get so much direct attention from a faculty member, and as with my time with Gerhard at Kew, I was going to make the most of it. Having someone hold my hand through the process for a change made my new knowledge feel pleasantly less hard-won than usual. So much of the learning I did back in Montreal was frustrating trial and error, or conferring with other grad students who knew only marginally more than I did. It wasn't the faculty members' fault—their workload was so high in administrative duties and grant writing that they rarely had time to spend in their own labs. I was vaguely aware that the monkey's paw–esque reward for being among the best scientists was not directly getting to do the actual science anymore. It was a fact I chose not to think too hard about, because I wasn't sure what it meant for the image I held of my career in research.

In the evenings with Pat's family, a quiet domestic scene unfolded. His fourteen-year-old daughter Sarah arrived home from school and worked away at her homework until Pat and his wife Donna, a chatty, welcoming woman who filled the role of botanical librarian at the Garden, could get supper on the table. Once everyone was seated, I had the privilege of watching the family of three reconnect after their day out in the world. Sarah was soft-spoken, happy, and intelligent. She had fallen in love with Korean TV dramas and spoke animatedly about one she had just started watching called *Secret Garden*. Donna explained with obvious affection that Sarah loved all things East Asian and planned to earn her bachelor's degree in Asian studies one day. That Donna could converse knowledgeably about Korean television stars with Sarah made it clear that she'd educated herself in order to share her daughter's pastime. Pat, for his part, listened intently and with a subtle smile playing at the corners of his mouth. They all seemed so warm and at ease with one another.

Parents of teenagers will no doubt find this scene unsurprising, but I hadn't been in the presence of a teenager since I'd finished being one myself. In Montreal, my domain was one of mid-twenty-somethings, that inflection point between youth and adulthood, in which some of my friends were recent parents adjusting to their new responsibilities and need for routine and order, while others were carefree partiers still out mining the city for new experiences. The former seemed to be fighting their way, red-eyed and irritable, through tantrums, dirty diapers, sticky fingers, and sleepless nights without much in the way of recompense. I lived among the latter and was nothing short of horrified by the lives of those who had crossed over.

Though nothing happened at Pat and Donna's house that wasn't happening every evening in millions of homes across the United States, the gentle warmth of the family's evening routine was my first window into what it looked like to have made it through the early years of parenthood and be reaping the rewards. They were close and happy. They enjoyed each other's company. I felt myself making the strange and subtle shift between identifying with the children in a family to identifying with the parents. Rather than looking at Sarah from the point of view of someone who had not so long ago *been* that age, I saw her as adults saw her. I could imagine myself with a child like this. Someone to talk to and share interests with. Someone whose way of seeing the world would broaden my own, as Sarah's interest in another culture seemed to do for her parents.

Tucked away in my warm bed in the basement at night, my mind drifted back to a different time. Is this what my teenage years might have looked like had my own family remained intact? My older brother had moved out when I was still quite young, and years alone at the farm with my father were happy, but in a quieter way than the carefree banter I heard at Pat's table. I felt a twinge of jealousy at the depth of relationship that even a family of three could create.

In a two-person home like the one I'd grown up in, there is only one relationship, however warm and affectionate it may be. In a two-parent, one-child home like Pat's, there are three, and those three generate a richer tapestry of interrelationship. In the stereotypical family of my grandparents' era, with two parents and three children, there are ten different relationships underpinning the household. Each new family member represents a wealth of new interactions and complexity. The mathematics of it are inescapable.

This made me wonder: Is parenthood a unique chance to change the past? An opportunity to experience what you missed as a child? Still, the disconnect between what I saw in Pat's home and what I saw my new-parent friends coping with was *vast*, and at home I was seeing firsthand what it could do to someone's career plans.

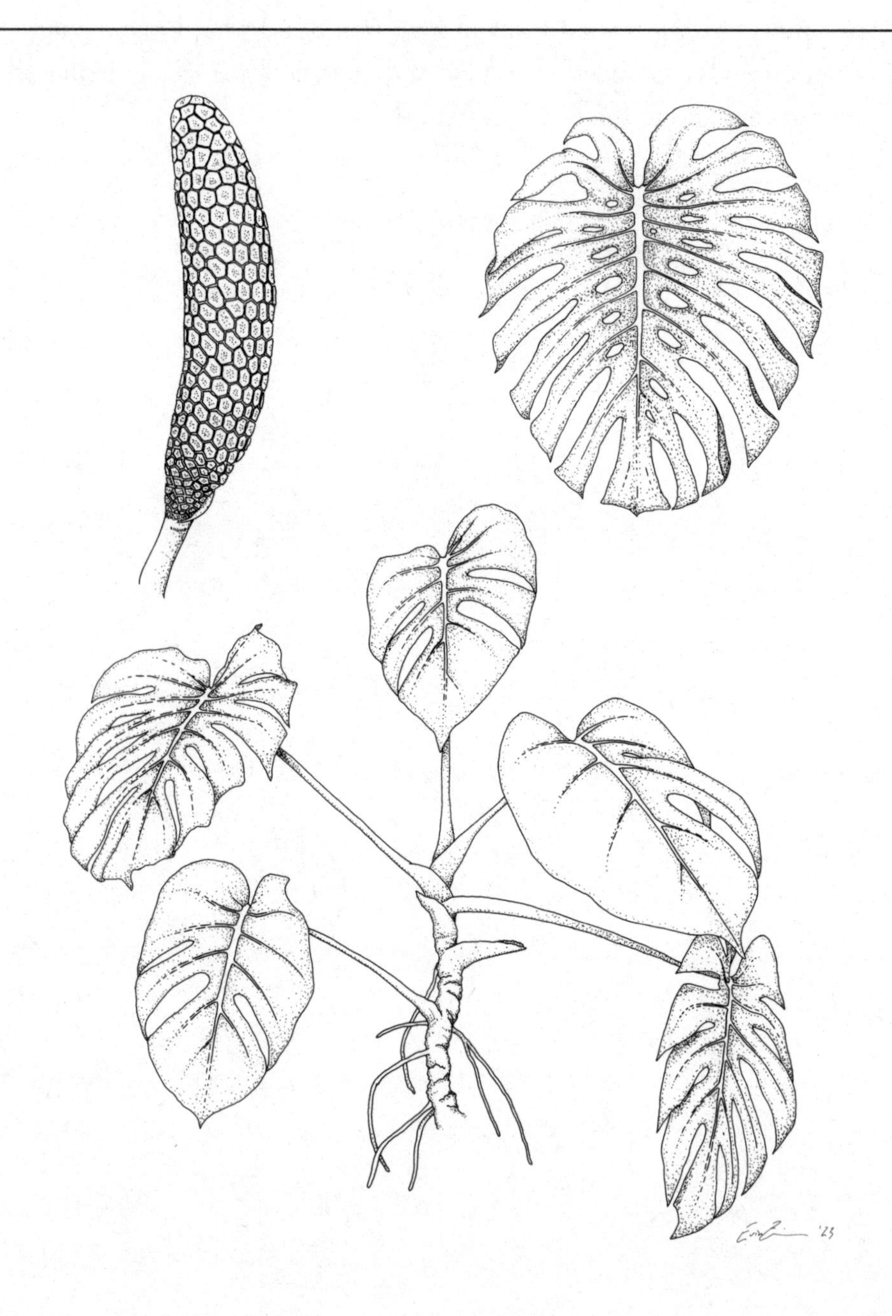

CHAPTER 3

TINY WORLDS

Our imagination is struck only by what is great; but the lover of natural philosophy should reflect equally on little things.

—Aimé Bonpland & Alexander von Humboldt, *Personal Narrative of Travels to the Equinoctial Regions of America, During the Year 1799–1804, Volume 2*

There are no more wonderful or casually disregarded creatures on this Earth than the plants. I don't simply mean in their utility to humans—though we wouldn't be here without them—but in the sheer diversity of form arising from the plant kingdom. Take something as straightforward as leaves. Their main role is to capture sunlight. So what's required here? A flat surface to intercept light—a vegetal solar panel. The simple leaves of something like a magnolia tree might seem both sufficient and inevitable. But what about a hotter climate, where the light is more intense, the air drier? Here, we find thicker, more leathery leaves, sometimes protected with wax to help keep the plant's water from evaporating, sometimes covered in fine hairs to keep the wind from pulling it away. In a colder climate, leaves are reduced to needles filled with antifreeze to withstand the weight and chill of the winter snow. An aquatic environment might call for highly divided, lacy leaves that cut down on water resistance, or air pockets that allow the leaf to float up toward the sunlight. And this is just leaves. Every part of a plant's body—roots, stems, vasculature, cell arrangement, and most of all, flowers—are subject to endless tweaking, each one a distinct set of evolutionary solutions to the challenges of its surroundings. Every one is unique. The poet William Blake saw the world in a grain of sand, but if you know how to look, there's one in each plant as well, an epic and agelong story that silently tells where it came from and what it took to survive in that place.

BACK IN MONTREAL, ONE OF the other PhD students in our lab group, Carole, had her first child. In a lab group of ambitious, motivated women, Carole was the one I secretly envied. Toned and athletic with chestnut hair and an intense blue-eyed gaze, she seemed to have energy and drive enough for anything. In the lab, she was knowledgeable, efficient, and unafraid to move out of her comfort zone learning new techniques. Evenings and weekends, when I collapsed exhausted into a heap and watched TV or read unchallenging books like Agatha Christie mysteries, she attacked competitive biking, learned to build and manage databases, and kept up on the latest scientific literature. She often seemed to have the favour of our shared advisor and was generally the person who knew everything. She was on track to finish her doctorate well before I finished my own. Carole was the person to whom I looked to remind myself that I could always do more and be better. I assumed that in having a child with her wife, also a former student of the institute, one who had moved into education after a master's degree, the child would be slotted neatly in among her plans for herself and little would change in her day-to-day work. Maybe she thought that, too.

What actually happened was that she vanished. Or so it seemed to me. When her son was born, we didn't see her for months. Eventually, she brought him in to meet us, but showed only a passing interest in what had been happening in the lab. She looked content, but tired in a way I'd never seen her before. When you spoke to her, she talked about the hard work of caring for a baby and about how much she adored her son, not about the relative merits of different methods of data analysis. Though she did eventually return and did eventually finish her PhD sometime after I did, the change seemed to be permanent. Competitiveness and ambition were never her overriding characteristics after that.

I felt both betrayed and self-righteous. Betrayed because the woman I'd placed on a pedestal had disappointed me—I had no one else with the same drive to hold myself up against and shame myself into working harder. Self-righteous because now I could be better than her by never letting motherhood distract me from my career. This is what happens when you have kids, I thought. Whatever mattered before gets left by the wayside. This is why women rarely make it to the top. Carole just didn't want it badly enough. Wasn't willing to sacrifice whatever was needed in order to be a scientist. I would never let this happen to me.

This story backed up my already entrenched narrative about motherhood, so I didn't really interrogate it. I just revelled in the petty thrill of getting to feel superior to someone I'd long felt less than, and carried on. I wonder now how many people looked at my retreating form with a similar mix of disappointment and smugness just a few years later.

My admiration of Carole was hardly surprising. At that time, Anne had stocked her lab exclusively with smart, highly motivated women, and I wasn't going to be the exception. Working in a lab full of women after having spent so many years among mostly men, first on the farm, and then in physics, and finally in Larry's lab, pulled me in two different directions. On one hand, it was inspiring; the concentration of ambition, ingenuity, and expertise was unrivalled anywhere else I've worked before or since. The seedy underbelly of that fact was that the arrangement seemed, against my will, to put my competitive streak into overdrive. These were my friends, but I could never seem to stop seeing them as the competition.

As the months went on, the attention I lavished on my plants, coupled with the isolation of my post-Eric existence at the institute, created a growing fixation on my work. I stayed late. I arrived early. I popped in on the weekend to redissect a flower I wasn't sure about. Away from work, thoughts of their shapes and structures played on a constant loop while I went through my day. I felt disconnected from the friends and family with whom I couldn't really discuss my work. My research became my life and my identity, while my hobbies receded into the background. The only thing I did for fun was to write a weekly biology blog about some unusual plant or animal. It made me feel like I'd accomplished something more tangible than a thesis that wouldn't be done for years yet, and gave me an excuse to broaden my knowledge of evolution under the guise of a pastime.

I spiralled inward. I'd seen this happen to plenty of other graduate students in my time at the institute. I simultaneously recognized that it wasn't healthy and welcomed what seemed like a new level of focus and dedication as I became the person I needed to be to pursue my career.

This utter devotion is considered the price of admission to a life in science, and burnout is worn like a badge of honour. Nothing else can matter as much. The history of science is filled with the stories of brilliant researchers—men, typically—obsessed with their work, devoting their lives, night and day, to the pursuit of truth and discovery. They are the heroes of our mythology. We rarely pause to ask who fed their children. Who rocked them and dried their tears and sacrificed the long days and years that it takes to nurture an adult human into the world. These heroes were not, by and large, childless men.

My experiences with Sarah and Carole showed me that you can look at a thing and see it in two very different ways. Was motherhood an exhausting, career-destroying slog, or the warm companionship and comfort of a family dinner? I didn't know. I was always just looking in from the outside.

A CAREER IN ACADEMIC RESEARCH looks very different today than it did even a generation ago. In theory, academia is the ultimate meritocracy—you advance according to your contributions and level of experience. At one time, completing doctoral research was enough to move on to a faculty position where, with further work and achievements, it was possible to be granted tenure, a permanent and protected position within a university. Far fewer doctorates were given out, the number bearing more similarity to the number of tenure-track positions available than it does currently. Many people embark on a career in science without realizing the extent to which this picture has changed.

Larry was right that there's always room for the best, but he may not have known how *few* of them. The straightforward and relatively stable careers that people of Larry's generation had, in which it's possible to go straight from graduate school to a tenure-track position and continue to do hands-on research throughout your career, no longer exist.

Today, the number of doctoral degrees awarded *massively* outpaces the number of permanent positions within academia. Degrees awarded have increased steeply over the past few decades, showing an average annual growth of 3 percent in the United States since the late 1950s,[1] with the total number of doctoral recipients more than doubling from 1985 to 2021.[2] Meanwhile, the number of faculty positions has changed little.[3] For that reason, rather than moving directly on to permanent research and teaching/faculty positions, young scientists in most fields must first become postdoctoral researchers, or postdocs. This tier is a sort of holding tank for those who finish their PhDs still determined to do academic research, allowing them to further sharpen their skills and learn to administer a lab and write grant proposals under the wing of a faculty supervisor. It is an apprenticeship that didn't used to exist but may now be where a researcher spends much of their twenties or thirties, working toward something permanent.

Surveys suggest nearly half of those finishing a doctoral program intend to follow it up with a postdoc.[4] In the life sciences, where the mismatch between PhDs and faculty jobs is particularly severe,[5] hopefuls can expect to spend around five years working as a postdoc,[6] versus a median of less than three for other surveyed scientific fields. Academic positions that do come open are barraged with high numbers of applicants,[7] and the number of research publications those who get hired have under their belt keeps rising,[8] shifting base expectations upward over time. In the end, though the majority of PhD graduates say that a career in academia is their first choice,[9] less than a quarter will eventually land a tenure-track position.[10]

ALONGSIDE MY LAB WORK, I spent much of the following year digging into the literature on current and previous legume systematics research. This would all lead up to what's called a comprehensive, or a qualifying exam—a several-hour-long oral exam in which I would need to prove I knew enough about my field to be considered a doctoral candidate. The exam consisted of two parts. The first was a detailed presentation on my project as a whole and how I'd carry out the experiments that needed to be done to test my hypothesis. A small panel of faculty members would question me to ensure I understood the theoretical basis of what I proposed to do and why exactly I was doing it. As a doctoral student, you're meant not only to be able to carry out experiments but to truly understand the thinking behind them. The second part was theory based. A month before the exam, I'd be given a complex and open-ended theoretical question on a topic related to my research and would need to prepare a second talk that comprehensively answered that question based on my reading. This would also be subject to grilling by the committee.

The PhD qualifying exam was a rite of passage that graduate students uniformly dreaded, and people did occasionally fail, though this was rare. It's hard to say whether it was rare because exam committees were reluctant to set someone back so severely in their studies by making them repeat the process or because the sort of people who did the qualifying exams were also the sort of people who would rather gouge their eyes out than fail a test.

Preparing for the exam meant months of even less human contact than research normally involves. As I sat alone in the library, digging back through decades of other scientists' work on our understanding of evolution, I had time to reflect on the recent departures of several friends who had once wanted a career in research but had eventually chosen to leave to pursue more easily attainable jobs in medicine or education. I had never given serious thought to what I'd do if I left research, but watching one colleague after another call it quits made me wonder why I thought I was any different, why I thought this would all work out for me, despite the odds.

Grad students in the sciences are nothing if not devoted sufferers of impostor syndrome, and I was no different. Nearly all of us, right up to the most chronic overachievers, believe we're one minor slip-up away from everyone discovering how stupid we actually are, and that we've (paradoxically) been cleverly faking it all this time. You look around, and everyone seems to know their stuff better than you do. Academia is full of people with unreal abilities to remember who authored a paper they read five years ago that made a point germane to the present conversation and to offhandedly say, "Yes, 'Smith & Jones 2011' addressed this exact problem and suggested that . . ." I've never had this kind of recall, which put me at a distinct disadvantage in the "sounding smart" competition that runs at all times and places in grad school, whether the competitors acknowledge it or not.

In recent years, there has started to be a greater focus on and awareness of the mental health toll grad school takes on its entrants, but at the time I was there, that hadn't quite happened yet. You did your best to cope and kept your poker face on. I'll never know how many of the other grad students I worked with each day were struggling, but the statistics suggest it was a lot of them.

We were living in a perfect storm of risk factors. Graduate programs are a distillation of smart, ambitious people who are used to being the best of a group. Finding yourself suddenly in a crowd where you're nothing special is a big shock to the ego and undermines your confidence at the exact time you need to be projecting it. The work is hard, the hours are long, and you're often far from, or at least too busy for, the friends and family who would help relieve the isolation and keep things in perspective. All this to say nothing of the constant fear that you're not going to get a job in the end . . . or not one that's worth the emotional and financial cost of what you've put yourself through to get it.

An international survey of 6,300 PhD students conducted by *Nature* in 2019 found that more than a third of respondents have sought help for anxiety or depression brought on by their studies.[11] Ranked top among their concerns were uncertainty over their job prospects following their degree and difficulty maintaining a good work-life balance. North Americans were also more likely than any other group globally to suffer from impostor syndrome, as I did. With over half of the students surveyed reporting that they worked more than fifty hours per week, it's hardly surprising they weren't happy with their work-life balance. The grad students that I worked with regularly spent evenings, weekends, and holidays working, either at the lab running experiments that needed checking at odd hours or at home, digging into the bottomless pile of reading that came with each narrow subfield of biology. It was a badge of honour to always be working, and expecting to knock off at five o'clock every day was a sign you were either lazy or, more charitably, didn't yet understand what was expected of you.

A life in science is a series of contradictions you must try to hold side by side. We pride ourselves on being smart, but secretly feel stupid. We work long hours, but worry it's never enough. We know a great deal about our fields, but years of study have also made us acutely aware of how little we *really* know. It's a constant pull between feeling proud and never feeling that you're good enough. Our profession has so many heroes held up on high, great minds that have changed the world, but most of us will work all our lives and fade into obscurity having never done anything of importance . . . and we know it. Striving for a career in science is trying to change the world and knowing you'll almost certainly fail.

I learned that year that I was being awarded an Alexander Graham Bell doctoral scholarship from the Natural Sciences and Engineering

Research Council, meaning the Canadian government had seen fit to throw over $100,000 at me over the first three years of my PhD research, presumably believing me to be a good bet as a future scientist.

In a way, it was the vote of confidence I'd needed, but at the same time, I was more uncertain than ever. How can anyone feel they deserve—and can deliver on—such a huge investment?

I had started out in Larry and Usher's lab washing glassware—a position colloquially known as the "dish bitch," regardless of the gender of the person doing it—but fairly quickly worked my way up to doing actual lab work. One of my first tasks was to use a hand cranked machine called a microtome to make very thin slices of the growing tips of various vines from the grape family so they could be inspected under the microscope. These plants form tendrils that wrap around whatever they come across in their surroundings, giving the vines support as they grow. Tendrils are what morphologists—scientists who study the external physical forms of plants—refer to as derived structures, meaning they evolved from some other part of the plant. Many groups of plants have evolved tendrils, and they have used different strategies to do so.[12] In some cases, leaves lengthened and thinned over millions of years, developing the ability to twine around objects. In others, it was a branch or even a flower stalk.

My thin slices of vine tips allowed us to view the individual cells that made up the tendrils as they formed, their shape and location giving us clues as to their origin. In the case of the grape family, the tendrils came from an evolution of the flower stalk, conscripted to give support and allow the plant to reach higher and intercept more light.[13] As they stretched to catch the sun over an unimaginable timespan, some of the stalks surrendered their blooms in favour of a new purpose. This is the endless shape-shifting of evolution.

In an age of quick, cheap gene sequencing and myriad technological advances in molecular biology and elsewhere, the practice of searching for understanding simply by looking very closely at the world, as I was doing with my vine samples, seems quaint. Natural history is often treated as an old-fashioned pastime held over from the Victorian era, when corseted ladies and tweed-clad gentlemen collected seashells and wielded butterfly nets. But so much of our foundational knowledge about the world has come to us this way. And indeed, in the case of questions like the origin of tendrils, it's the only practical way to get those answers.

Today the gathering of natural history–based knowledge is being increasingly set aside in favor of experimental and model-based approaches. Though we can and have learned a great deal from a handful of fully gene-sequenced model organisms such as fruit flies—or botany's analogue, *Arabidopsis*, a small, fast-growing plant in the mustard family—the fact is, we still know little to nothing about most species of life on this planet.

A survey of early-career biologists published in 2016 found that an overwhelming majority saw the importance of natural history to our understanding of life, but very few felt confident that they knew enough to teach a course on the subject.[14] When courses aren't taught, knowledge isn't transmitted to the next generation of students, and in time, no one is doing the fundamental work of building basic knowledge.

Natural history is how we know our world. It asks, "What species is this? Where can it be found? How many are there? How does it survive? What is it related to?" This approach underpins every other life science because until we know the basics of an organism, and those that influence it, we can't ask further questions, test theories, understand patterns in nature, or predict the effect of change on an environment. And most pertinent in our current age, we can't mitigate that change.

In its earliest days as a formal science in the eighteenth century, natural history represented an effort by its practitioners to impose some sort of logical system on the seemingly endless and overwhelming diversity of life they found at home and in their travels. As Europeans made their expansionist push across the globe, new plant and animal species were pouring into their museums and universities at a rate of thousands each year. To create order, they collected, named, described, illustrated, and classified.[15]

The age of exploration led to the rise of the explorer-naturalists beginning in the Enlightenment. A mutualistic relationship formed between naturalists who sought to collect and classify ever-greater swaths of the living world and imperial powers that needed to buttress their economic power with new and valuable foods, medicines, and luxury items from their overseas colonies. Plant collecting became an obsession of the Western world as the prospect of both money and fame beckoned to naturalists willing and even eager to risk their lives on long and arduous voyages to discover new species.

Collectors were often naval officers or ships' physicians. Sometimes they were sent to far-flung locales as part of a colonial government and took the opportunity to study the place. One of the seventeenth century's greatest collectors, Sir Hans Sloane, whose vast collection would later form the core of London's Natural History Museum, was sent to Jamaica for fifteen months to act as physician to the governor.[16] In that short period, he collected over 1,500 specimens and enough knowledge to write a two-volume natural history of the island, introducing Europeans to hot chocolate in the process. Others set out with no goal other than to collect and document whatever new flora and fauna could be found in a new frontier.

The scientific expeditions and collections of the eighteenth and nineteenth centuries form the basis for much of the work that's been done in the life sciences since then.[17] The price for adventure, discovery, and possible scientific fame was steep, but the payoff in knowledge was enormous. Specimens that survived long trips home under less-than-ideal conditions could, once reaching their final home in a herbarium or a museum, last for centuries. For botanists in Europe, the ability to directly compare local plants with those from distant shores started to uncover a bigger picture of how life on Earth worked. Realizing that similar species could be found on continents separated by entire oceans led people to think that the world map perhaps hadn't always looked the way it does today. Comparing fossils to species still living added weight to the idea that species can change over time, and that the planet might be far older than the dominant Christian thinking suggested.

Collected specimens were the individual clues that helped pull naturalists out of a theology-based mindset and move them into the evidence-based science that's carried us to where we are today. In two and a half centuries, natural history has produced a logical system of naming and classification that serves as a framework for incorporating new discoveries, a theory of evolution that became the major unifying force behind all life sciences, and the knowledge base upon which all other fields of biology are built. Faded leaves and flowers pressed on paper may seem antiquated and unimportant, but each is a tiny piece in the foundation of a singularly huge and stable structure of knowledge.

SOUTH AMERICA HAS ALWAYS FEATURED heavily in the annals of European botanical exploration. Joseph Banks, before becoming an important figure at Kew Gardens, visited Brazil as part of James

Cook's famed Endeavour voyage in 1768. The purpose of the voyage was to observe the transit of Venus from Tahiti as part of an international scientific effort to improve estimates of the distance between the Earth and the sun and to search for an undiscovered southern continent (which famously became Europe's first contact with Australia). After sailing from England, they stopped off in Rio de Janeiro to restock supplies and make repairs to the ship. Though Banks passed through only briefly, his time there produced the naming and first scientific description of a now-popular horticultural plant: bougainvillea. In total, the Endeavour voyage brought back enough new plant specimens to increase the number of plant species known to Europeans by around a quarter.

When Prussian naturalist and polymath Alexander von Humboldt and his companion, French botanist Aimé Bonpland, landed on the Venezuelan coast in 1799 and began to document and collect the plant life they found there, Humboldt wrote that they ran around "like fools" and would go mad if the wonders of the place didn't soon let up.[18] Their five-year expedition through the region would net over *sixty thousand* plant specimens comprising six thousand species, two thousand of which were new to Europeans. Of equal importance, the expedition set Humboldt on the path to his worldview, centuries ahead of its time, that the natural world should be viewed as an interdependent whole that humanity has the capacity to fundamentally disrupt, anticipating the modern science of ecology.

Humboldt's *Personal Narrative* of his expedition served as a source of inspiration to Charles Darwin as he travelled along the continent's coast and to points inland on the voyage that would one day help him piece together his theory of natural selection. Darwin's recollections of the voyage of the *Beagle* in turn inspired Alfred Russel Wallace, eventually to be Darwin's codiscoverer of natural selection, to travel to the

Amazon valley to gather evidence in his own search for the mechanism of evolution.[19] Though Darwin and Wallace would further develop their ideas at home in England and in southeast Asia, respectively, both were originally inspired by what they found in South America. Such is the muse-like magic of that continent, as I'd learned firsthand in my own quest to understand the evolution of my plants.

CHAPTER 4

INTREPID ENOUGH

Hence, a traveller should be a botanist, for in all views plants form the chief embellishment.

The scene . . . was novel, and a little danger, like salt to meat, gives it a relish.

—CHARLES DARWIN, *Voyage of the Beagle*

It was our third week on expedition in the Guyanese rainforest, and we'd hiked out to the base of a small but fast-moving waterfall to see what plants we could collect around the wet rock face. Getting to it involved scrambling over a series of large boulders with long drops to one side. It would have been a nerve-wracking climb at the best of times, but as our party got about a third of the way up, the sky opened and started to pour in that abrupt fashion particular to the tropics. The mossy rocks instantly became slippery to the touch, and it wasn't clear whether it was less dangerous to keep going or try descending back the way we'd come.

I looked at Karen, our expedition leader, battling her own wet boulder up ahead. She seemed unruffled, as always, but very focussed.

"Just go slowly and take your time," she called back through the rain. "We'll get there."

So we kept going, and managed to get to the rock face to make some collections, including a few delicate orchids clinging to the stone. But it was obvious we weren't going to get back down the way we'd come. One of the field assistants who had stayed below circled around and cut a path through the dense forest near the top of the falls. It was an alternate way out, but we still had to get up to the head of the trail he'd cut. That meant climbing a near-vertical wall of wet cut-grass, a plant named for its hand-slicing capacity. I pressed myself into it, grabbed the cut-grass with my bare hands, and tried to scramble up. Any time I stopped scrambling for a moment I started to backslide, slipping down the grassy wall toward the huge boulders and rushing water below.

After a few minutes of frenzied activity, I peeked my head over the top of the wall. As someone with a profound snake phobia, I was considering that the ill-timed appearance of one right now would be the end of me. The universe must have laughed at that thought, because just then, what looked like a terrier-sized rat poked its face out of the bushes inches in front of me. I was so startled that I nearly lost my grip and plummeted to a grisly fate below. Some deeply buried instinct for survival surfaced just in time. It took the entirety of my willpower to remain motionless in front of what turned out to be a creature the Guyanese call a labba: *Cuniculus paca*, a harmless South American rodent that can reach up to twenty-five pounds and, I'm told, tastes delicious.

Making my way back to our field site after the climb, feeling the exhaustion that follows in the wake of a flood of adrenaline, it occurred to me that I'd just risked my life for the sake of some pressed plants.

It was May of 2009, and we were a team of two American botanists, a local flora expert, two field assistants, a local guide and observer from the Amerindian community on whose land we were collecting specimens, and me, representing the University of Montreal as a freshly minted plant systematics PhD student.

We were in Guyana to conduct a general sampling of the flora of an understudied region of the Guiana Shield, a billion-year-old geological formation and the inspiration for Sir Arthur Conan Doyle's *Lost World* that runs along the northeastern edge of South America. That meant we'd be collecting anything and everything, from tiny clumps of moss right up to samples from the tallest trees of the canopy. I was specifically there to learn more about legumes in the field; how to identify them, where to look for them, and how to make proper collections of them, as well as to try to find any Dialiinae species I could. The Dialiinae are spread across the world's tropics, but a handful of species in several different genera grow in Guyana, and any fresh material I could get my hands on would be a boon.

It was the kind of expedition, in search of new life and new understanding of a place, that botanists and other naturalists have been engaging in for centuries as they pushed out across the map to find what was out there and bring it home to try to make sense of it. Of course, nothing is new to the people who have always lived there.

WHEN ANNE HAD APPROACHED ME with her usual understated grin and told me I'd be accompanying a month-and-a-half-long botanical expedition to a remote part of the South American rainforest, where I'd sleep in hammocks each night and climb mountains every day, and where the snakes could kill me, my first reaction was excitement. My second, a fraction of a second later, was abject terror at remembering that I am not, in fact, a rugged adventurer. I say she "told" me rather than "asked" me because there was no question necessary—no one in my position would dream of turning down this kind of offer. The expedition was the opportunity of a lifetime, but was I the right person to go? Would I be able to stand up to the physical punishment and hard labour of tropical fieldwork? I had my doubts. But I wanted so much to be that person. I'd read about the adventures of the great explorer-naturalists since childhood, and nothing seemed more exciting than finding myself somewhere incredibly remote and beautiful, seeing something that other people would never see and doing the work of documenting life. I dreamed of herbarium specimens with my name on the collector line. And in my wildest dreams, I got to name a species myself, with some clever and beautiful Latin name that would last beyond my time in this world.

"You'll be a great adventurer, finding amazing flowers and living with monkeys," Eric said brightly when I ran up to his lab to tell him. Moving into his frequent role as morale booster in my life, he pushed me to embrace the sense of excitement I'd felt initially. "Imagine all the new plants and beautiful places you'll see." I countered with the fact that we'd be wandering among both very large and lethally venomous snakes.

I'd been terrified of snakes since early childhood, beset by recurring nightmares of finding them in my bed with me. I'd wake up absolutely convinced that they were writhing unseen around my feet and would launch myself out of bed, too afraid to get back in. Even as an adult I had these dreams, and would have to talk sternly to myself about dreams versus reality in order to keep my feet under the covers. Sometimes I'd still have to check.

He casually brushed this off as he continued pipetting various reagents. "I'm sure they'll be terrified of you. Just sing while you walk!" A typically optimistic Eric reply.

Some of my fear must have gotten through to him. Months later on the morning I left, as I finished my last-minute luggage checks near the door of our little rented duplex in Pointe-Saint-Charles, emotions stretching taut between excitement and apprehension, he slipped an envelope into my hands.

"Something to help you feel better. A letter. Don't open it until you need it," he said, and hugged me. I tucked the letter into a Ziploc bag to keep it dry, slipped it in next to my passport in my carry-on, and headed for the airport.

DUE TO THE DIFFICULTY OF moving over land in a tropical forest, we planned the trip along the path of the Mazaruni and Kako rivers, east of the Venezuelan border in the Pakaraima Mountains. Despite the pristine feel of the surrounding forest, the larger rivers in the area had been contaminated with mercury from gold mining operations, so we didn't bathe or wash food there, though this became less of a concern as we moved upriver and away from populated areas. Our boats were long, wooden, brightly painted yellow and blue, with outboard motors that struggled to move their heavy cargo at a crawl. But travelling by river allowed us to put a good amount of distance between the camps that we set up. Each of five camps was home for anywhere from a couple of days to a week as we sampled the unique plants in the vicinity. The camps consisted of a small, sandy clearing near the water, created anew at each site by cutting away undergrowth and chopping down any thin, spindly trees that got in the way.

In tropical forests, soil fertility is low and light competition intense, so nearly all the trees are very tall and thin, appearing younger than they really are. Our Guyanese companions then bound the cut trees with durable, woody vines into a tarp-covered framework with astounding speed and artistry. The result was a small, open-sided structure within which we could hang our hammocks and be protected from rain and large falling fruits—a real concern in the rainforest, where there are hard, fist-sized fruits that could knock you out if you were unlucky enough to block their downward path. Another cleared area nearby, left uncovered, served as a cooking and work space. We even had a big, straight stick with shortened branches coming off it on all sides, like a miniature coat rack, that we pushed into the ground and hung our plastic coffee mugs on.

Each morning, we rose around 5:00 a.m. and got ready for our day, quickly downing a breakfast of oatmeal and sweet coffee with evaporated milk. On collecting days, we'd set out early, well before the heat of the day set in, slowly making our way through the undergrowth, cutting paths with machetes. Each day we hiked a new area, which could be a mountainside, a swamp, a valley . . . you name it. In that region of Guyana, the mountains are a flat-topped variety called *tepuis*, or tabletop mountains, unique to that part of the world. They jut abruptly up through the jungle in isolation, rather than as a continuous range, like a series of ancient tree stumps grown over with moss, making each one home to a group of unique indigenous plant species that evolved alone on an island thrust up toward the sky.

Depending on the abundance of a given plant, we'd take anywhere from one to ten samples of it, unless the plant was potentially rare or threatened, and then it was left alone—better it remain unknown but alive.

We'd move slowly over the land, collecting whatever we found to be sorted later. If there was time, I'd sketch some of the more interesting finds in the notebook I carried. As the hike went on, our big canvas collecting bags got heavier and heavier.

Occasionally, we had to climb a mountainside shortly after a heavy rain. This was one of the most challenging activities of our fieldwork because the soil was loose and sandy, and it was easy to lose your footing and slide backward suddenly. My natural reflex when this happened was to reach out blindly and grab a tree trunk for support, but this was something I had to train myself not to do because falling was often preferable. The plants there were better armed than you might expect, like the well-named "monkey no-climb" tree, *Hura crepitans*, that sports a trunk covered in stout, inch-long thorns that could easily puncture skin. Some other trees were covered in long, thin, needlelike spikes. Many had no defense of their own, but there'd be a caterpillar perched on the bark whose hollow, glassy hairs were filled with a chemical concoction that caused severe pain and blistering. In still others, it was a scorpion. Sharp eyes and deliberate movements could be the difference between disaster and a near miss. One day, we returned from one of these climbs and I noticed a strange, persistent pain in my calf. I pulled up a pant leg and drew a thin, inch-long needle from one of the trees directly out of my flesh. How it got there without my noticing I never knew.

A week into our six-week stint in Guyana, after an afternoon collecting on a punishingly hot swath of open savanna full of browned grasses and the odd, alarming snake skeleton, I got sunstroke and spent the evening vomiting and running off into the trees to dig new latrine holes with a machete. I was weak from sickness, and my whole body ached. Everyone took a "camp day" or two at some point in the expedition, staying behind to rest while the rest of the group set out, but I was the only one who needed one our first week out. From their suddenly curt and skeptical manner toward me, I worried the others were already wondering if my coming along was a mistake. I spent the day drawing plants in my field journal, trying to lose myself in a sci-fi novel, and willing myself not to poke at the cold ball of dread in my stomach telling me this trip was going to be a painful, embarrassing disaster.

I had hoped I'd be tougher than this. More like our expedition leader, Karen Redden, an indomitable woman who steamrolled through any difficulty, physical, administrative, or psychological; she attacked mountains with relish and still had the sense of humour left to make lewd jokes with the Guyanese men at the end of the day. Karen was a research associate and collector for the Smithsonian Institution's Biodiversity of the Guiana Shield program.[1] Over a career that included a dozen collecting expeditions to the Guiana Shield region, she encountered venomous snakes, skirted lethal drops, and contracted malaria, leishmaniasis, *and* anthrax. None of it put her off fieldwork. She'd wanted to be a field biologist since childhood. "I loved looking at old herbarium collections and imagining the collectors trudging through the jungles," she told me. "You can't help but daydream about their adventures."

She was the image of the botanical adventurer I was supposed to be, and I was disappointing her.

Lying in my hammock that night, I pulled out Eric's letter. I hadn't intended to read it so soon, but my moment of need came a little earlier than expected.

> *My intrepid Erin,* it began. Ha ha. I wish.
>
> *You are strong, brave, and enterprising.*
>
> *Don't be worried. Why worry in paradise, in the Kingdom of the Plants?*
>
> *You'll be gone only a short time. Too brief for you, probably.*
>
> . . .
>
> *Each time you want to see me, imagine I'm still waiting for you at the airport, waiting for your plane, waiting to see your face, waiting for our summer to begin.*
>
> *I love you,*
>
> *Eric*

It soothed something in me that had been jangled by that rough and relentless place. I fell asleep willing myself to be more the person my partner thought I was.

BOTANICAL COLLECTORS WORKING IN THE tropics have always had to endure hardship. Being exhausted, sore, and bug bitten and missing your loved ones may never change, but my scientific forebears had it worse. I was away from home for all of six weeks, but expeditions in the eighteenth and nineteenth centuries typically ran several *years*. Months spent travelling by boat presented a trial before the fieldwork even began. Darwin suffered horribly from seasickness, which left him confined to his quarters writing letters home to family with sentiments like "I hate every wave of the ocean" and "I loathe, I abhor the sea and all ships which sail on it."[2] Humboldt before him endured the punishing heat of crossing the sun-scorched plains of the Llanos, where he and Bonpland stuffed their hats with leaves to try to keep their heads cooler, and were burned badly by the sun.[3] The same expedition found the two men half frozen at seventeen thousand feet above sea level on Ecuador's inactive volcano Chimborazo. Dressed inadequately for the conditions, they suffered numb and painful extremities and even bloody feet when the rocks sliced up the soles of their shoes.

Malaria and other tropical diseases were also sources of misery for explorer-naturalists. Alfred Russel Wallace nearly died of malaria while exploring the upper Rio Negro in Brazil in the early 1850s. His younger brother, Herbert, joined him and died in a yellow fever outbreak. Richard Spruce was gravely sickened by malaria, suffering from recurring bouts of illness for the rest of his life.[4] His most lasting contribution as a botanist was, ironically, to help increase cultivation of the cinchona tree, from whence the treatment for malaria comes, at a time when overharvesting was rushing it toward extinction.

AS IT WAS WHEN THE explorers of the past visited the forests of the neotropics, malaria is still a concern combatted with bed nets and pharmaceuticals. The antimalarial drugs I was taking brought on strange, vivid dreams. For some reason, perhaps a counterpoint to my untethered life in the jungle, mine all seemed to swirl around some distant domestic future. In one, Eric asked me to marry him, and I happily said yes, then was struck with panic and a feeling of being trapped once I stopped to consider what I'd promised. In the real world, back home in Ontario, my whole family was gathering for a younger cousin's wedding, and my mind kept coming back to all the fun they were having without me, and how people are rewarded for following the path that's expected of them. Going to graduate school and studying obscure plants wasn't ever going to be something my family at large understood or celebrated me for, as most graduate students of esoteric subjects inevitably find out.

In another dream, I suddenly had an infant daughter. In that way of dreams where you don't question what's led up to your present situation, I walked into the kitchen of the farmhouse I grew up in and set down a car seat with her bundled inside. Everyone was happy for me, but I was suddenly struck by the terrifying realization that I had no idea what to do with her, or where my life would go from here. Again, I felt panicky and trapped. Several times in Guyana, I woke up crying or breathing hard in my hammock. Once, Karen told me, I fought like I was trying to escape, and she worried I'd fall out. Each time, I'd lie awake and listen to the stillness, the insect song and distant animal cries of the nighttime rainforest. There is nowhere like the jungle at night to remind you that the path of your life hasn't yet narrowed, and anything is still possible.

I was at a point in my life when there was a growing divergence between the life events a person my age would usually experience and what my life in research looked like. Often, they didn't seem compatible. Children, and maybe even marriage, felt like constraints on the kind of life I was chasing. My brief brush with the mother-daughter bond hadn't left much of an impression, and I just didn't see how the personal sacrifice and, as I saw it, loss of dignity that comes with raising messy, chaotic children could possibly be worth it. Motherhood was the stuff of drug-fuelled nightmares for me.

A TEST OF HOW WELL I was adapting to the rigours of life in the rainforest came a few weeks into the expedition. We were collecting in a dense swath of forest and found it opened onto a large swamp. It was the first such area we had come across, and it promised a variety of new specimens waiting to be collected. Karen and Ken, the other botanist, as well as Delph and Timo, the field assistants, had set out into the reed-filled, calf-high water to see just what interesting plants were to be found there. Knowing how common water snakes were in tropical swamps, my phobia and I opted to stay by the edge of the water and catalogue what I could find there. I quickly lost sight of everyone, their voices eventually trailing away behind them. I collected for a time, but my teammates were gone for what ended up being hours.

I sat down on a fallen tree a few feet from the water and started making notes in my field journal to pass the time. The hours kept stretching out, and I'd finally decided to work up the nerve to follow my team when Delph suddenly appeared from one side along the bank, heading straight for me, walking quietly but very quickly. He looked afraid—not something you wanted to see on the face of someone hired for his stoicism and experience in the bush. Swooping by me almost without slowing, he caught me firmly by the arm and told me to follow him, *quietly.* At a walk that approached a run, we moved into the forest. We were well out of sight of the water before he finally slowed and turned to me. Catching his breath and gradually resuming the calm demeanour which, until then, had been his defining feature, he described the enormous anaconda that was lying motionless, stretched out among the grasses at the edge of the water, just a few feet from where I had been sitting.

Lying in my hammock that night, a small collection of facts on the natural history of anacondas played on repeat in my head. The green anaconda, *Eunectes murinus*, is one of two species of anaconda that is found in Guyana; there are four species globally. It has a maximum verified length of thirty feet, sometimes weighing over four hundred pounds.[5] It is the largest snake in the world. Feared throughout human history and often killed on sight, anacondas have a reputation as merciless man-eaters, crushing the bones of their victims before swallowing them whole. Sleep didn't come easily, tired as I was.

I later learned that the truth about anacondas is somewhat less dramatic than the stories. Attacks on humans are vanishingly rare: anacondas prefer to flee rather than risk pursuing unfamiliar prey of unknown strength. Neither do these huge snakes crush their victims' bones. Their constriction blocks the circulatory system from delivering blood to the prey's brain,[6] although aquatic attacks are just as likely to result in drowning. Further, while there have been attacks, there is no *documented* case of an anaconda killing a human.[7] Sitting there together, each apex predators in our own right, I was theoretically as much a threat to the snake as it was to me. Nevertheless, I was happy not to have learned what it's like to have a giant snake try to drown me in a swamp.

It wasn't *all* hardship, of course. There were beautiful times, too. Scenes of such perfection they washed away the aches and filth and exhaustion for a moment. A couple of weeks into the trip, we were hiking up the side of a mountain, Guyanese field assistants Delph and Timo up ahead cutting through the undergrowth while I and the other botanists, as well as the Guyanese flora expert, Claudius, followed and collected as we went. We reached a summit and the land flattened out, but the forest was incredibly thick, and we couldn't see much of anything through it.

We walked across it for what seemed like forever as I silently dreaded how long it would later take to get back down. A roar of water came into earshot, though it wasn't clear where from.

Suddenly, it was as if we were spit out by the forest and I found myself standing on an open plateau, looking down an enormous, vertiginous drop. I wasn't close enough to fall, but my mind hadn't been ready for the vast plunge of open space. We had come to a wide, flat expanse of rock over which the water flowed thinly before gathering into a huge, multitiered waterfall with a drop of hundreds of metres ending in a misty set of rapids far below. Guyana is famous for its waterfalls and is home to the largest single-drop waterfall in the world, Kaieteur Falls, which is about four times higher than Niagara Falls.

The whole scene was framed in the bright green of the forest. The sun was shining, and the sky was a bright robin's-egg blue dotted with the fluffy white clouds that I always associate with South America. It was stunningly beautiful. I heard the others laughing and splashing around as the mood lightened with the air. We were so high up that we could drink the water straight from the stream without fear of pollution from the mining projects far below.

"How's this for a view?" Karen asked, walking up beside me as I sat on the rocks, basking in the sun. "Happy you decided to come along?"

As we sat, sipping that cool, clear water, immersed in all that beauty, I thought to myself about how I'd have run away from this trip a hundred times before now, how I'd almost chosen to stay at camp that very morning.

Things were starting to get easier. My body ached less. My feet had stopped blistering and bleeding. Some days I could be a part of the jokes. I hadn't stopped missing home, but sometimes I could go an afternoon without thinking about it.

❧

ONE OF THE MORE GLAMOROUSLY risky activities of botanical collection is tree climbing. Trees and the air plants and woody vines that live on them make up more than 60 percent of the plant species in neotropical forests like the one we were in.[8] The highest canopy trees of a wet tropical forest can reach over sixty metres, and they don't have many branches on the way up because those branches wouldn't reach the sun and would therefore be a nutritional liability. That means you have to get yourself *way* up to sample a branch with leaves or flowers. Climbing spikes are large, crescent-shaped pieces of metal that strap onto your shoes. They have teeth positioned on the bottom such that if you nestle the crescent around the circumference of the tree's trunk and step down, the teeth will catch in the bark firmly enough to bear your weight. Once you're up there, your colleagues on the ground hand you what amounts to a set of pruning shears on a very long pole, operated by pulling an attached rope. At the moment you cut a sample, you're holding a heavy pole in one hand, a rope in the other, and are attached to the tree by both feet and a rope around your waist, the ground far below.

Experienced climbers will scuttle up trees with impressive speed, giving you the impression that it isn't really that hard. Learning to do it for the first time is another matter. First, there is the natural tensing of the human body when your brain realizes it's at a height from which a fall would hurt. Then there's how complicated the heavy spike actually is to position in order to grab the tree properly, using only your foot and without a clear view of what you're doing. Do it incorrectly and you'll slip; you won't fall, because your other foot-claw will still be dug into the bark, but you'll probably scrape your face on the tree and maybe twist your other ankle. What's more, once you reach a certain height, you're on your own figuring out how to do the whole process in reverse to get down. Then there's the possibility of wasps and stinging ants defending the tree. If you get well above the ground and bumble into the nest of some angry, swarming insects, you are in a *lot* of trouble. You can't jump, and it's hard to calmly and correctly place your feet to climb down while under attack from biting and stinging insects. But for many of the trees in a wet tropical forest, climbing is the only way to sample. So we climbed.

Inexperienced as I was, I never got far enough up one of the big trees to sample anything, so I remained on ground support whenever there was high-altitude collecting to be done. But climbing up to even a modest height and seeing the forest floor from above, surrounded by the leaves of small trees and lianas in that airy green lightness, was one of the moments when a life in science paid off in rare views.

Though my expedition concluded with no major injuries or mishaps, it doesn't always turn out that way. On her next expedition to Guyana, in 2012, Karen developed an infection in her hand while out in the forest. The emergency antibiotics she always carried in the field failed to stop it, and she became feverish, her hand swelling dangerously. It began to turn blue, and even cutting into it with a razor blade to relieve the pressure didn't help. Knowing she was in serious trouble, Karen hiked from dawn until dusk with the assistance of a colleague to return to a small mining village as quickly as possible.

Once there, she went into septic shock. She barely remembers taking a supply plane back to the capital city and having emergency surgery—essentially having her hand flayed open—to try to clear the infection. Though the surgery probably saved her life, it didn't cure the antibiotic-resistant infection, and she was airlifted back to New York to undergo further surgery to try to save her hand. She worked in Washington, D.C. at the time, and quipped to the surgeon to please try to save her middle finger: "I'm a D.C. driver, I need that finger."

This is the dry sense of humour you need to be a fearless botanical explorer. Even so, she spent a week in the hospital and took two months to fully recover. She still lacks sensation in much of her hand due to nerve damage.

THE EVENINGS IN THE FIELD were no less tiring than the days were. One of us prepared a quick supper over our little gas stove. Most days, we ate beans or lentils cooked with meagre spices in a pressure cooker and mixed with soggy rice boiled in too much water. Everything reeked of kerosene. On a whim, I had brought along a single small bottle of chipotle-mango flavoured seasoning. This little luxury was seized upon by one and all and came to be known as The Spice, à la Frank Herbert's *Dune*, to be enjoyed with practically every meal. To me, the rainforests of Guyana will always taste like mango and chipotle.

Night closed in quickly so close to the equator, but all the day's collections had to be processed before the heat and humidity destroyed them. For a northerner, short days coupled with intense heat felt fundamentally wrong and disorienting. Hot days should be much longer, and the dusk always took me by surprise. In the failing daylight, and then by LED headlamps and battery-powered lanterns, we sat on upturned five-gallon buckets, hunched over a mess of leaves and newspaper, back muscles screaming. Reggae played on a battery-operated MP3 player someone brought along. A lot of the songs had raunchy lyrics, and the field assistants laughed and made dirty jokes. Just behind this, we were serenaded by the haunting, echoing sounds of distant howler monkeys that always sound-off as the day ends, earning their Guyanese name, "six o'clock monkeys." When the wind blew, it would trigger the exploding seed dispersal mechanisms of certain fruits, and we'd hear them popping in surround sound out in the firefly-spotted darkness of the forest. It was otherworldly.

Processing the plants meant photographing them beneath a tripod-mounted camera and removing a leaf or two to be placed to dry in silica for later DNA extraction. Flowers, if there were any, were popped into a phial of spirits to preserve delicate structures for study.

The tree branches and small, herbaceous plants—carefully dug up to keep their roots intact—were then laid out on sheets of newspaper marked with the collector's initials and collection number and then pressed in stacks and tied into bundles. These bundles of newspaper were placed in garbage bags and soaked in cheap alcohol we'd acquired from a rum distillery in Georgetown, the capital city of Guyana, for the remainder of the trip. This would keep fungi and bacteria at bay, though with an ugliness and a chemical stench that belied the delicate beauty trapped inside.

There's an art to how plant specimens are laid out. The alcohol renders them inflexible by the time they're unpacked again, so they need to be arranged in much the way they will eventually appear on their herbarium voucher. This means the leaves are splayed out as much as possible to not overlap, and at least one leaf must be turned so that the underside is visible for study. This is because even microscopic details of the underside of the leaf, such as the arrangement of stomata—tiny holes for gas exchange—or the minute, hairlike trichomes, can be helpful in identifying the plant and evaluating its relationship to other species. Each collection, even if it's the same species as another but found in a different location, is marked on its newsprint in waterproof grease pencil with a different collector number.

Collector numbers are one of the little-known and peculiar dramas of the botanical world. Each collector has a system. They vary, but most use their initials followed by a number that indicates the sequential order in which the collection was made. The bigger the number, the more collections they've racked up during their career. This has led to a situation in which serious collectors will compete to see who can achieve the highest numbers, with some well into the tens of thousands. As with most unpoliced incentives, it's also led to a bit of questionable behaviour, with some collectors choosing specimens that didn't necessarily need to be collected again or having a number of other people—students, for example—collecting under their name. I've heard these people referred to only half jokingly as "number whores." All in all, though, the competition has probably benefitted natural history collections greatly.

BACK IN THE HOT BUSTLE of Georgetown after weeks in the jungle, the rhythm of our work shifted abruptly. Long, slow days of hard physical labour were replaced with checklists and frenetic runs around the city on errands to make sure we accomplished all we needed to before our planned flights home. There were permits to be obtained, checks of our collections to be done at the University of Guyana, field books to be copied and given to the botany department there, and mountains of paperwork. We hired a driver, K.P., friendly and talkative, to take us on all these errands, and it was often he and I driving around the city together while Karen and Ken did paperwork back at the small, single-story rental house that served as our base. I enjoyed these drives because K.P. would happily give running commentary on the people and places that we passed, providing a local's insight into the workings of what felt like a large town. He also knew where to stop for the best fresh fruit, regional delights like mamey apple, sapodillas, and Suriname cherries that rarely make their way out of the tropics, and certainly never as far as Canada.

Georgetown is the most vibrant city I have ever visited. It is crushingly hot, much more so than in the rainforest, but everyone seems to be hurrying on their way somewhere, dressed in beautiful, brightly coloured clothing. Traffic is chaotic. There are few stop signs, and drivers seem to communicate in a local dialect of long, short, aggressive, and friendly honks that I could only barely make sense of. Everyone is honking. Animals are everywhere, from the ubiquitous small, ownerless dogs, which also seem to be off on important business, to the horses and cows tied up in people's small front yards, to the lizards, cockroaches, and frogs that are on everything and occasionally jump on you. Driving through it, windows open, the smell of the city veers between the stench of raw sewage in open drainage ditches at the sides of the roads, to the delicious aroma of roti being cooked street side, and fresh fruit in stands here and there along the way. Georgetown is a glorious assault on the senses, and I loved it.

At our rental-house-*cum*-workspace, there was a rush of activity. Every one of our hundreds of collections had to be pressed in a new sheet of newspaper to keep it from getting damp or moldy. The plan had been to use the large plant dryers at the university to do the final drying needed to ship our plants back to the United States, but they weren't working, so Karen, ever resourceful, jury-rigged a dryer out of some space heaters, a tarp, and a few other odds and ends, and set it up right there in the house. It was probably an egregious fire hazard, but it did the job.

Each plant, once fully dried, was ready to be mounted, labelled, and distributed to a herbarium. In any case where only a single collection of a plant could be made, it always went to the University of Guyana, since it's their flora. Any further samples were distributed to the foremost herbaria of the world: the Smithsonian, which was partially funding our trip; Kew Gardens; New York; Paris; the Netherlands; and elsewhere. Researchers all over the world could then use them to investigate scientific questions about the plants and ecosystems that exist on the Guiana Shield—and how they may be changing compared to earlier expeditions.

More so than in the jungle, where I had struggled to keep up, Georgetown was a place where I could contribute. There are moments in scientific research that are glamorous and romantic, like stepping out onto the top of a stunning waterfall and collecting exotic plants. But there are many, *many* more times when the work of science comes down to having a detail-oriented mind and a very high tolerance for tedious, repetitive tasks. This is where I shine. Our last days in Guyana, I organized and cross-referenced months of receipts, I re-newspapered hundreds of specimens; I coordinated errands and got things where they needed to go. As hard as I tried out in the bush, I think I was more helpful in three days in Georgetown than during weeks in the forest. It made me realize, with equal feelings of disappointment and hope, that my place in science might not be as an intrepid adventurer, but as an organized mind pressing order onto chaos.

Fieldwork, and collections in particular, forms a foundation for research on biodiversity, species distribution patterns, global and local extinctions, invasive species, and the effects of climate change and other types of environmental degradation. Even if we do already have a good idea of what's present in a certain location—which isn't the case in so many parts of the world—collecting specimens provides a snapshot of that place at a single point in time, allowing us to see how it changes. Collections are priceless in their way because we can never go back and take that same snapshot again. At its root, it's all based in comparison: What's here now? What was here before? What's here but not there? Was it there before? Is there more of this now, or less?

If anything, collections become more valuable over time as science progresses and new methods of analysis are developed. We can now analyze DNA from species that went extinct decades before we knew what DNA *was*—but only if we have the collection in the first place. That's why the current trend of decreased funding for collecting is so concerning. A career as a pure collector is nearly unsustainable now, in part because while very specific scientific questions may get funding, broad, exploratory work doesn't. Just a few years after our expedition, Karen, a modern-day explorer-naturalist if ever there was one, left her career as a field biologist, in large part because of the poor pay and constant struggle for funding. "You can't just be a collector anymore," she told me sadly. I might have asked myself at the time what it meant for someone like me if someone like *her* couldn't make it work.

Short funding cycles discourage the kind of long-term observation and collecting needed to really understand biodiversity and how it's changing in a particular environment. Meanwhile, museums that maintain natural history collections are dealing with reduced budgets and the elimination of staff positions.[9] A lack of funding and staffing means that specimens—samples someone risked their safety to collect—now sometimes sit in storage for many years, unidentified and unstudied, because there is no one available with the necessary expertise, or the few that have it are too overworked. Most institutions have cabinets full of collections waiting to be identified so they can be further studied. A study published in 2010 found that fewer than a fifth of new species are examined and described within five years of being collected, and projected that *more than half* of all "undiscovered" plant species are sitting in herbaria right now . . . just waiting to be discovered by someone able to understand what's sitting in front of them.[10]

After a month and a half in Guyana, I was exhausted and sore, but also more at peace with the experience than I'd thought I would be. I was eager to get home and rediscover the luxury of a real bed, fresh vegetables, and hot water. I wasn't as brave or as strong as I'd hoped, but I was proud and a bit awed at having helped to find and collect plants that will be available for scientists to study for decades and even centuries to come. For a moment, I was part of a long line of people who cared enough about knowing the world to take big risks. It made the hard work and the danger feel worthwhile. Our expedition brought in over seven hundred collections belonging to ninety different families of plants, some of which may prove to be new to science.

CHAPTER 5

LEARNING TO SEE

> I have never done anything else since, oil-painting being a vice like dram-drinking, almost impossible to leave off once it gets possession of one.
>
> —MARIANNE NORTH

It's a still and silent early Saturday morning at the herbarium in the Montreal Botanical Garden. I'm sitting at a black-topped workbench next to a floor-to-ceiling window that overlooks the rose beds. The mildly warm day is glowing with the green gold of late April as life really gets going out in the gardens. I can see visitors strolling along the walkways outside, but not many at this hour. I'm beginning the final year of my PhD, and I can sense things starting to wind down. My morphological work is finished, my molecular work nearly so, despite many months of struggling with old and unworkable samples of rare plants from which I can't draw any good DNA. I'll soon be moving into a long stretch of data analysis and thesis writing. There are more days behind than ahead of me in my time as a doctoral student now, and I'm starting to hear a small voice in the back of my head urging me to think about what's next. Eric is also a year from graduation, and I know that voice is in his head, too. Wherever we end up, I know it won't be as simple and sheltered as our lives are at this moment. Our days as students are nearly done.

The bench at which I sit, adjacent to the compactors that store the specimens, is bright with the morning light. I've been coming on the weekend because in this part of the building, at least, I know there will be no one else. I can spread my books and tools out as much as I wish and lose myself in the work. It's a sort of work that is particularly easy to lose myself and my mornings in—I'm drawing my plants. As with my morphological work, I've got herbarium specimens and a microscope set out, but now I'm trying to channel what I'm seeing through a pencil as I look. It's harder than I thought it would be, but the focus it requires brings a trancelike calm, and hours vanish without my noticing. It's only when the glow of the early morning peaks and then drains away that I realize how long it's been.

Studying the morphology of my specimens demands that I look carefully, taking stock of many individual features of the plants and noting them—illustrating those same specimens, to my surprise, requires something else—drawing requires that I truly *see* them. In their entirety. The difference sounds trivial, but it isn't. The curves and textures can't be recorded on a score sheet. The places where the shadows fall along the raised veins of a leaf aren't captured by words. But illustration requires me to absorb and understand these complex features. The shift from looking to *seeing* in a holistic sense isn't an easy one. We want to draw what we think is there, our idea of the thing, and not necessarily what our eyes truly perceive. Our minds want to clean things up, to schematize the world in order to aid our understanding, but nature is messy. Life is messy. There are irregularities in leaves, asymmetries. Insect damage and bruises on petals. The different parts of the plant come together in a more complex way than our brains would like to imagine. These are the elements I have to force myself to see and accept when I draw. They are the topologies that make the hours vanish.

This noticing, this continuous flow of attention when our minds would prefer discrete chunks, is where the true value of plant drawing lies, and it is why the shapes of my trees will be etched into my memory forever.

BOTANICAL ILLUSTRATION HAS EXISTED IN some form since prehistory, first appearing as cave paintings showing the uses and dangers of different kinds of plants in our environments.[1] In the Western world as far back as Aristotle, the ancient Greeks used hand-drawn illustrations in herbals—books that discuss plants' medicinal properties. But they were very different from the lifelike depictions of today; the drawings of the period were stylized rather than realistic. Then, as intellectual pursuits stagnated during the Middle Ages, botanical illustrations from the ancients were handed down unquestioned, copied and recopied by scribes with no botanical training, rendering them less and less accurate over time. As such, the practice of botanical illustration declined over the centuries. That changed during the Renaissance, when a renewed interest in knowing the world through direct observation, as well as in naturalistic rather than stylized portrayals of living things, brought about a dramatic improvement. It was at this time that some of the first modern-looking illustrations were produced, combining scientific precision and beautiful composition.

As a reproducible way of conveying detailed information about plants to a large group of people, scientific illustration had to await the development of the printing press in the mid-fifteenth century. With this, botanical books became much more widely available, though they often still used illustrations copied from ancient manuscripts. Realistic portrayals of plants in books became more common in the sixteenth century, when both growing and painting flowers became popular. In the late 1500s, this led to the creation of florilegia—books that depicted flowers ornamentally rather than based on their uses by humans. The books contained little text but showed the plants accurately, often accompanied by associated birds or insects. Florilegia stayed popular well into the seventeenth century and were made possible due to evolving methods of image reproduction, which by then included engraving and etching on metal plates.

Botanical illustration didn't fully come into its own until the so-called Age of Discovery, when Europeans undertook the exploration of the Americas for reasons of both science and greed, and accurate plant depiction became indispensable. Collections that could not be brought back alive, and that would lose important features such as colour when dried, needed to be illustrated so that they could be accurately described and classified.[2] Though early expeditions were often privately financed and undertaken by enthusiastic amateurs, by the latter half of the eighteenth century they were largely organized and carried out by governments. These voyages employed professional botanical artists and set a higher standard of scientific accuracy. In fact, the illustrations of the late eighteenth and nineteenth centuries reached a level of accuracy only to be matched by the adoption of photography in the twentieth century. Many of the most prominent explorer-naturalists of the time were talented illustrators in their own right, including proto-ecologist Alexander von Humboldt, natural selection codiscoverer Alfred Russel Wallace, and the zoologist Ernst Haeckel, famous for both his peerless artwork of marine invertebrates and for accusations of racism and scientific fraud. The most notable absence is Charles Darwin, who was known to have little to no artistic talent at all.

An established style coalesced during the eighteenth century in which the plant is pictured against a white background, divorced from its natural surroundings,[3] mirroring the appearance of a mounted herbarium specimen. Around it, details of the flowers and their component organs, as well as the fruit, are often pictured to help with identification. Though the plant itself is faithfully reproduced, it is removed from its environment and the other living things with which it interacts. This may reflect the absence of interconnected ecological thinking at the time the style was developed, but it persists even now. How ironic that even as near-perfect accuracy in botanical illustration was achieved through centuries of improvement, context within the living world was lost. Still, as we'll see, a few artists and illustrators such as Maria Sibylla Merian and Marianne North are known for showing the landscapes and animals entwined in the lives of their subject plants.

In a time of cheap, high-quality photography when cameras, alone or mounted on microscopes, can capture images on any scale and at almost any angle we wish, it's not always clear why it might be worth the time and expense of having a trained illustrator draw a plant for a scientific publication or educational material. But cameras can only capture an object exactly as it appears, and counterintuitively, that isn't always what's needed. Photography lacks the ability to selectively emphasize the most important features used to identify a species. Colourless trichomes, for example, are tiny, hairlike structures that are barely discernible in photographs but can be drawn fully visible in illustrations. Or take the more ephemeral elements of a plant. For example, stipules, which are tiny, leaflike bits attached below the base of a leaf, are clues to a plant's identity, but are usually shed very early in the growing season. An illustration can show them as they briefly appear on the plant.

Those are features that exist but can't easily be seen. Illustrations can also show us what can never be seen because it no longer exists. Morphological studies on both living plants and fossils can be used to infer the appearance of extinct plant species, in much the same way we infer the appearance of dinosaurs. This can even be done speculatively for plants for which we have no fossils at all but that the evidence tells us probably existed and looked a certain way. Like dinosaurs, these inferred plants are nothing more than a collection of presumed features until they are put together by an illustrator to create a vision of how it might have appeared as a whole. This is something photography can never do.

For the botany novice, drawing is unparalleled as a learning tool.[4] Studies have shown that when students, whether adults or children, are encouraged to draw a plant they engage in a detailed study of the plant that not only improves their observation skills but also develops their awareness of its structure and how that relates to function. *They actually see how it works.* Drawings tend to be more information rich than written descriptions and allow newbies to skip over the intimidating array of technical terms involved in describing plants.

Humans learn to recognize complex objects holistically—we recognize an oak tree from a distance because it "looks like an oak tree." Drawing, which necessitates understanding how different parts are connected, forces a holistic view of the subject, as opposed to writing, which facilitates the individual description of discrete parts of the plant. So those who have drawn a certain species tend to remember it better. What's more, most learners just *enjoy* drawing more than writing as a learning tool, and that enjoyment can lead to a greater engagement with and interest in science and the natural world. Illustration can be what brings the "green blur" most people see when they look at plants into focus.

And if right now you're saying to yourself, "But I can't draw to save my life," it's worth noting that studies also indicated that students tend to underestimate their ability to effectively convey information this way. You may be better than you think.

FOR ALL OF THE ADVANTAGES botanical illustration presents, as a grad student, it can still be a hard sell to an advisor who needs work to be published and projects to progress at a timely rate. I was lucky. Not only did Anne see the value in her students beyond the hours of work they could put in, she saw their value beyond the skills needed specifically for the lab. Knowing that I could draw, she allowed me to make it a part of my work and take the considerable time required to do it. She met me where I was. The illustrations I produced were unlikely to benefit her, but she could see that they might be of value to me. I had started drawing simply to help me to understand the morphology of the plants, since the seeing required to draw is more than sufficient for noting those characteristics. But the hours spent bent over a microscope with a pencil and sketchbook at hand gave me a much greater sense of peace and satisfaction than pipetting and reading DNA sequences ever would. When anxiety for my career and sadness for the future of a warming world where many of my beloved Dialiinae species were in danger gathered like shadows around the edges of my mind, the drawings comforted me. Creating something beautiful can shore us up even in the face of intractable problems. I decided, with Anne's blessing, that I would illustrate all the research that I published from my doctoral work. Drawing would officially be a part of my scientific work, a privilege afforded to few in our modern era of fast-paced, intensive publishing.

Even now, when I read scientific research papers full of the names of plants I've never seen and no illustrations or photos, my eyes glaze over. Without images, the names are just lyrical scraps of Latin with nothing to attach them to. But give me a picture, and I'll care. I'll want to know how these singular pieces of the living world came to be the way they are, and how they do what they do. That was what I wanted for my work: that readers see the species I'd come to know so well, and perhaps to understand better why they mattered. I wanted to share with the larger world, most of whom will never see any of these obscure, jungle-dwelling giants, why my "babies" were so unique and wonderful and different from the rest of their family group. As any of us hopes for our children.

That summer, the annual conference of the Botanical Society of America took place in New Orleans, and our lab was attending. It had already been a few years since my time in Guyana, and I was ready for some intense heat and sunshine of the sort that Montreal summers just don't provide. Several years without much travel focussing solely on my lab work had made me antsy, and I couldn't wait for the vibrant street life I knew New Orleans would hold. Eric would be coming along on the trip for a bit of vacation time; he'd wander the city during the day while I attended talks, and we could spend the evenings and a day or two postconference together before going home.

On the work front, with the end of my PhD research drawing closer, it was time to start putting out feelers for what I'd be doing next. In New Orleans, and at a conference I was to attend later that summer in British Columbia, I hoped to talk with some researchers doing natural history work—morphologists, taxonomists, systematists—and ascertain whether they would be looking for a postdoc soon. If so, I was going to do my damnedest to make myself look like the best choice for the job.

Setting down in the Big Easy in late July, New Orleans didn't disappoint, in heat or in vibrancy. In sweltering humidity that lasted well into the evening, Eric, my lab mates, and I wandered Bourbon Street soaking up the flavours, from some of the best seafood we'd ever had to the novelty of walking around in public with a cocktail in a to-go cup. Walking into a quiet lecture hall at the conference centre the next day to listen to a staid academic talk was quite a contrast from the nonstop party taking place outside. I'm sure I wasn't the only person to walk in with a lingering hangover that the French refer to as *un dur lendemain de veille*—literally "a hard day after the night before."

Between talks on herbarium digitization and fern reproduction, during coffee breaks and evening drinks, I made the rounds of the natural history researchers, most of whom I'd met through Anne's introductions, to talk about my research, ask about theirs, and gently probe the topic of their upcoming postdoc needs. It had all the awkwardness of trying to figure out if someone you've just met is single so you can then circle around to asking them out on a date. As subtle as I tried to be, this is a common dance in science, and they could easily see what I was getting at. A theme started to emerge in the answers I heard.

"I love morphological work and would like to do more, but there's just no funding for it these days. I'm moving in another direction."

"I've had to move into doing pure molecular work these days, because it's what I can get funding for. And it's hard to get any money for hiring postdocs at all."

"I've shifted into ecology now, and just do ontogeny work as a sort of side hobby, since I can't get it funded. It's not something I could bring on a postdoc to do."

All were apologetic, and all had left off doing organismal work because research needs funding, and for that type of project, it wasn't there. Running in parallel was a trend of having no money to pay postdocs at all. It was a tier that had to be passed through on one's way to being a fully-fledged academic researcher, but it was becoming a massive bottleneck. It was easier to fund graduate students, who aren't as expensive since they're still students. Most aren't raising families and can take lower pay for their work, so for researchers running cash-strapped labs, their cost is more bearable. There were certainly postdoctoral positions to be had, but increasingly, it seemed like finding one in my field was going to be tough. Once again I wondered if this was a party I'd arrived at just a little bit too late.

A HIGHLIGHT OF THE CONFERENCE, after all the drawing I'd been doing back home, was a workshop being offered on botanical illustration. It was being given by Alice Tangerini, the longtime staff illustrator for the Smithsonian Institution's Department of Botany, and she would be instructing a small group on the techniques she used to produce her illustrations. I fell all over myself making sure I signed up fast enough to get one of the limited spots. On the day, about fifteen of us gathered in a small conference room and settled around a few long tables. Tangerini, a slender, soft-spoken, dark-haired woman in her sixties, showed us a few of the illustrations she'd produced as part of her work. Made with technical pens and India ink on a translucent, plasticky-feeling sheet called vellum, the drawings were sharp, graceful, and incredibly detailed. The sweep of a branch filled most of the large sheet, with close-up views of flowers and fruit filling the space around it, until the whole surface was filled with the lines and curves of one species, showing all its important parts. I ached to create such beautiful depictions of my plants.

We were given some images of different parts of a pawpaw tree, *Asimina triloba*, along with the tools we'd need to create the illustration. Tangerini showed us how to first draw, or in some cases trace, the parts of the plant in pencil on pieces of acetate that could be arranged aesthetically on a large sheet of Bristol board. Then we would carefully lay the vellum over the arranged drawings and reproduce them in India ink on the vellum. Once the basic outlines were complete, we could add shading and detail with dots or strokes of the pen.

I was instantly in love with pen and ink as a method of illustrating plants. The ink was unforgiving of mistakes, but the lines were so sharp and crisp. When Tangerini later had our pieces of vellum scanned and reduced so we could see how they would look if they were shrunken down for publication, I was astounded by how detailed and precise everything appeared. If a drawing looks good in a large format, it'll look *great* once you've shrunk it because the sharpness remains, but any little mistakes become much harder to discern.

I worked away for several hours that felt much shorter, trying to create depth and curves with only dots and lines of ink. When I stepped back, I was pleased. It was far from perfect, but there was an immediately recognizable *Asimina triloba* on the page. With a few minutes left in the workshop and most of the other participants still intently working away on their drawings, I approached Tangerini and asked her about her work as an illustrator for botanists. A tiny voice at the back of my mind suggested that if I fell short of my dream of a career in research, perhaps one in botanical illustration would allow me to stay close to the people and plants I cared so much for. As badly as I wanted a life in research, I knew I ought to be aware of what other possibilities might be out there. She seemed pleased at my interest and enthusiasm but circumspect about recommending her line of work.

"Well, there's not much call for it these days, I'm afraid," she said. "There aren't many of us left. Most are retiring and not being replaced. If you like to draw, you might be better to focus on a different subject."

And then the line that gave me déjà vu after my earlier talks with taxonomists and morphologists at the conference: "It's hard for the scientists to get funding for it."

I was crestfallen. It seemed like everything I loved most in science belonged to an earlier time.

WHEN I SPOKE TO TANGERINI again recently, nine years after our initial meeting, she was able to give a more nuanced picture of botanical illustration's place in modern science. When she began work in the late 1960s, the landscape for botanical illustrators like her was very different.

"There were illustrators associated with the main botanical institutions everywhere. In the States, we had ones at New York, Missouri, California, Texas, St. Louis, Florida," she said. "They were usually connected with herbaria because scientific illustrators were hired to draw from collections that were being described by taxonomists looking for new species or drafting monographs or floras." Monographs are in-depth taxonomic descriptions of a single group of plants, such as a genus or family, while floras are illustrated guides to the plant life of a given area, like a country or a region. Technical illustrators like Tangerini would work with the scientists to produce illustrations that emphasized the important or diagnostic features of the plants being used. The illustrations were done in pen and ink because printing in colour was expensive, whereas black-and-white line drawings were cheap to reproduce. The style followed a very long tradition of representing plants this way. "Basically up until digital [artwork], I would say even our twentieth-century drawings were all based on the line drawing that was done in the early herbals from the 1500s. They were basically outlines; everything was outline. It's the shape of your plant, and then also the surface features."

Though described by others as an artist, Tangerini seems unsure whether her elegant, detailed drawings constitute works of art. "I am drawing for a very specific audience, but at the same time, I'm still an artist at heart," she told me. "When scientists say, 'It doesn't have to look aesthetically pretty or pleasing,' I say, 'It has to be aesthetically attractive to the eye in order for the audience to even look at it,' whether the audience is a scientist or just a layperson." She paused. "It's art in the service of science. So it's still a combination of the two, because you're trying to bring your artistic ability to show this object for the scientist." She said that for her, the deciding factor is in the audience the work speaks to. "I've always said it's more the intention of how the drawing is made. Is the drawing made for a scientist? Is it made to be a scientific illustration?

And maybe it elevates itself to the point of being, later on, just considered a botanical artwork. Which to me, I think that's an elevation. Like it went from just being appreciated by the scientific community to being appreciated by the art community."

I asked Tangerini if, like me, she feels like she really knows and remembers the plants she's taken the time to illustrate. "I tend to know the plants that I've drawn because I have to look at them so closely," she said. "I have to understand [them]. If it's a group I've never worked on, I have to learn where it is in its family and what makes it look like that. And when I'm drawing different species, I'm asking, 'How do I show that it's something completely different?'"

But the work of a botanical illustrator is changing. In 2012, a rule change made it possible to publish a new species description in a digital-only format. This, coupled with digital-only scientific journals—called e-journals—becoming much more common in recent years, has removed the economic restraint of using only black-and-white drawings. Botanical illustrations can now be drawn digitally and have colour without incurring extra cost. But at least for taxonomic work, this new freedom of medium and colour hasn't led to more illustrations being produced. Tangerini points to the rise of molecular and genetic work as part of the cause. "They're doing more molecular work, and that has totally affected the job market," she said. "I've noticed the drop in the number of botanical illustrations in our journals, and also the escalation of using graphics that depict molecular data or evolutionary genomics. It's a totally different way of looking at plants rather than taxonomically, where you have an illustrator draw." With this type of work, she adds, "you don't need a drawing, you just need software to generate your findings."

Because pen and ink line drawings have been used in botanical illustration for such a very long time, they facilitate comparison of different species through the drawings. "The reason we keep to certain media in drawing has to do with how a scientist has looked at it in the past—if it's always been drawings in pen and ink. Visually, you want to keep a consistency in the appearance of drawings. You want to keep it to black-and-white line because they're comparing it to much older drawings in black-and-white line. Nobody seems to think about that. We talk to people in the lab, and they're like, 'No, it's all lab work in the future. It's all molecular. It's going that way.' But I see taxonomy as always being necessary. I see that visual comparison as being important."

For Tangerini, the tangible nature of her work is part of its value. "Illustrations—the pictorial depiction of a plant—is so important. Because it's there. The living plant will not always be there."

Today, women dominate the field of botanical illustration, but that's a status quo that's been in place for hundreds of years now. Like botany more generally, botanical illustration was considered during the eighteenth and early nineteenth centuries to be a suitably "ladylike" hobby for middle- and upper-class women to engage in. And like most activities undertaken primarily by women, it wasn't highly valued. Girls were taught watercolours alongside a basic education in botany, though they were not expected to use either to earn money. Some helped husbands, brothers, and fathers in their botanical research and publishing ventures, and were frequently uncredited for their work. Others, like Elizabeth Blackwell, used their illustration skills to earn a living at a time when there existed relatively few employment options for women.

Born in Scotland in 1700, Blackwell used her skill at botanical illustration to support herself and pay her husband's debts while he was in prison for running an illegal print shop.[5] Learning of a need for a new herbal, she drew the plants of the Chelsea Physic Garden with the support of several prominent botanists of the time, thanks to her family connections. She later engraved and hand coloured the work herself, producing a book entitled *A Curious Herbal.* It was the first complete herbal reference of the time and was a commercial success. Her ability to support herself in this way was particularly important as her husband couldn't seem to stay out of trouble, later moving to Sweden and being executed for high treason after apparently being involved in a plot to alter the Swedish line of succession.[6] As Ann Shteir points out in her book *Cultivating Women, Cultivating Science: Flora's Daughters and Botany in England 1760–1860*, Blackwell's herbal was one of the earliest botanical publications authored by a woman, and existed as a link between a new tradition of women as botanical illustrators, and an older tradition of women as herbalists.

Still other women used their skills in botanical illustration as a means to travel the world. Maria Sibylla Merian, born in 1647, was raised in an artistic household and encouraged in drawing and watercolours. She developed a particular interest in insects and the plants they depend on for food, and published in 1697 a book that she had both written and produced the artwork for on the topic of caterpillars and their food, as well as several earlier books on European insects that were illustrated with her engravings. Her work was admired for its accuracy and for the novelty of showing an animal alongside its food source. After nearly two decades of marriage, Merian left her husband to join a religious sect with her mother and two daughters. In her fifties, she felt compelled after seeing a collection of insects from Suriname to go there and study the insects in their native habitat. In a move that was unheard of for a woman at the turn of the eighteenth century, she sold off many of her belongings and used the funds to spend the next two years observing and drawing plants and insects *in situ* there. Merian's works are still appreciated today for their detail and elegance, as well as their nod to the future science of ecology, showing the interdependence of organisms and the various life stages of the insects portrayed.

As the nineteenth century progressed, botanical illustration became one of few paying professions accessible to women, and came with the tacit understanding that their participation was limited to producing artwork rather than scientific involvement. Women were now employed as botanical artists in such numbers that many publications were dependent on their work, though they were still frequently uncredited for it. British women who found themselves in a privileged position due to reasons of wealth and class were able to use colonial connections to move through the world with relative ease for someone of their sex, painting the local flora wherever they went.

Probably the best example of this is Marianne North, who was born in Sussex in 1830 to a wealthy family and a politician father. Like Merian, North's career took off relatively late in life. Having promised her late mother to always stay with her father, she travelled with him in Europe and the Middle East until his death in 1869. Once her father had passed away, a grieving North, now age forty, was free to visit the tropics as she had long wished to do. For the next fifteen years of her life, she travelled abroad, painting flowers and their surroundings in a style that was bright and vivid and more artistic than scientific in leaning. Being from a wealthy, well-connected family, North was received by colonial administrators and local bigwigs wherever she went.

Though her paintings weren't detailed enough to be useful from a taxonomic perspective, she was encouraged by Joseph Hooker, then director of Kew Gardens and a prominent botanist of the time, to paint as much as she could as a means of capturing novel and increasingly rare flowers: "wonders of the vegetable kingdom" that "are already disappearing, or are doomed shortly to disappear, before the axe and the forest fires, the plough and the flock, of the ever-advancing settler or colonist. Such scenes," he wrote, "can never be renewed by nature, nor when once effaced can they be pictured to the mind's eye, except by means of such records as this lady has presented to us, and to posterity . . ."[7] As Alice Tangerini said to me, the living plant will not always be there.

More than eight hundred of North's paintings eventually came to rest at Kew Gardens, in a gallery she helped finance herself and which is, to this day, believed to be the only permanent solo exhibition by a female painter in the United Kingdom.[8] Almost weekly during my time at Kew, I'd wandered among her bright, vivid paintings, hung floor to ceiling in the gallery, and imagined what it must have been like to have such a wildly adventurous life.

In addition to her paintings, North collected specimens for Hooker, dipping her toe into involvement in the scientific study of plants at the time. North's travels began in 1871, which is noteworthy because they happened amid a period of rapid professionalization in British botany, when women and amateurs were being pushed out in a bid for a greater appearance of respectability in the field. Still, North's paintings and collections led to the description of plants new to science,[9] and a genus, *Northia*, as well as four species, were eventually named in her honour.

ONCE THE CONFERENCE HAD ENDED and New Orleans's sudden concentration of botanists had dissipated, Eric and I had another day and a half to roam the city before flying back to Montreal. We meandered aimlessly through a city we didn't know, wanting to see what was beyond the flash and noise of Bourbon Street. In the leaden heat of midsummer, we followed the curve of the Mississippi westward as far as Audubon Park, captivated by the intensity of greenery possible in a climate so much hotter and brighter than the Canadian summers we were used to. We wandered among bald cypresses, magnolias, and live oaks draped in Spanish moss and resurrection fern that made them look as though they were melting in the sun.

We'd been living in a sort of shared daydream together since the previous winter. I'd awoken in our apartment on the edge of the ice-choked St. Lawrence River on a bright, frigid New Year's Day to Eric bringing me a breakfast tray in bed. Under a cover he lifted off dramatically was a large pancake covered in blueberries. Contrary to his plan, my tired and slightly hungover brain hadn't recognized that the blueberries actually spelled something, so he had to ask me to look carefully before I dug in. When I eventually managed to resolve the words "Will You Marry Me" in the berries, I was speechless. I truly hadn't expected it, focussed as I was on my research and where my career might be headed. I was so intent on planning my future, I didn't see that future arriving. He lifted up a small, upturned bowl that I also hadn't noticed on the tray to reveal a little bell-shaped silver ring box. Opening it, he knelt next to the bed and voiced his question, just in case I hadn't fully absorbed what the blueberries were asking me.

Like many of the turning points in life, it was over in an instant, but from that moment, we were off. We could now plan the years ahead knowing that we'd be doing it all together. The knowledge added bounds to a set of possibilities that at times had seemed almost *too* broad. Whatever we did, we'd need to make it work for both of us.

All this was how we came to be standing in a cramped yet charming old antique shop on Magazine Street in New Orleans as we made our way back to our hotel on our final day there. I'd gone in looking for teacups or charming oddities to take home as souvenirs and instead found myself bent over the counter staring at a pair of old wedding rings. The matching pair, which had come from a recent estate sale, were covered in grape leaves wrought in green and rose gold and felt like the perfect set for a pair of wayward botanists. I loved that they had once belonged to another couple who had presumably lived out their lives together. It felt like good luck. The shopkeeper, a New Orleanian of many generations, struck up a conversation with us, and we told her our stories and of our plans to marry. She was delighted to learn we were "plant people" and invited us to come and see her home on the bayou with its unique plant life the next time we were in New Orleans.

I flew home from Louisiana clutching a carry-on containing our two botanically themed wedding rings and thinking about how perfect they were as a representation of what I wanted my future to be—love and leaves. A happy marriage and the time and space to fill my days with plants. Excitement over the wedding we were planning for the following year and the prospect of getting to defend my thesis allowed me, for the moment, to push my failure to make progress on a postdoc out of my thoughts.

But as I worked through the final year of my PhD research, finishing up some last, difficult-to-capture DNA sequences, analyzing my data, and of course illustrating my plants, worry rooted ever more deeply in my mind. Drawing, in particular, though relaxing, was also a way of distracting myself from the fact that the end of my doctoral work was rushing toward me and I still had no plan for what came next.

I'd been scrolling through academic job listings and society bulletins as well as continuing to gently prod colleagues, but I wasn't having much luck. Whether because Anne was too tied up in her own demanding career path at the time or simply expected me to find my own way at that point, my next steps were never something we sat down and discussed together, and I felt like I was flailing. Though there were plenty of postings, particularly in the United States, there seemed to be nothing in my area of research, and I wasn't sure I was willing to relocate to a different country to work in a new field for only the year or two that a postdoc contract might last. The nature of Eric's work as an optometrist, not to mention the student debt he'd had to take on, meant that he could move anywhere in the Unites States or Canada—but only once. Then he had to settle down and start earning money. He was willing to go anywhere I wanted, but it couldn't be a series of regular moves over the next decade, which is what a lot of people pursuing a research career are forced to endure to eventually get that permanent job. The whole situation started to seem daunting. And under it all, bubbling up from the pit of my stomach, was the homesickness I'd never stopped feeling since leaving my rural birthplace. I missed the open spaces and endless stretches of green.

I missed being so in tune with the natural and agricultural cycles there that I could walk outside and *smell* what month it was. I missed home.

I still had nearly a year to sort all this out. Something would come up. I tamped the worry back down and kept drawing, the act of creation allowing me to exist only in the present for a little longer.

Later that summer, I flew to the southern interior of British Columbia for what would be my last scientific conference as a PhD student. Kamloops is a small city nestled in the Thompson Valley at the convergence of two branches of the Thompson River. Sitting in the rain shadow of the Coast Mountains, summers there are arid and sunny, with big, open skies that sweep away to the surrounding hills and grasslands spotted with bunchgrass and Ponderosa pines. After the mugginess of Montreal and the downright soupiness of New Orleans, the bright, dry heat of Kamloops felt fresh and light and invigorating. The conference was the annual meeting of the Canadian Botanical Association, a much smaller and more intimate gathering than the one in New Orleans, with many more familiar faces and only around one hundred people in all.

Despite its size, there was plenty going on. In addition to the usual scientific talks, there was a field trip to a nearby provincial park called Wells Gray, which holds a gorgeous water feature, Helmcken Falls, that we'd hike to while checking out the local flora. I was also accepting an award for the floral development work I'd done at Kew and exhibiting some of my illustrations at a small botanical artwork show the society was putting on. The conference felt like the perfect capstone to my doctoral work, encompassing the people, the science, the artwork—all the things I'd loved most about my time in grad school. Yet, for all the excitement, the occasion was bittersweet—I didn't know what my future held, so I didn't know when or even *if* I would see these people again.

It was dawning on me that if you leave research, you're also out of a social club. Only those involved in research can be members of scientific societies, so when your career is on uncertain footing, it's very easy to find yourself on the outside looking in, a fact that makes it all the more difficult to separate your identity and friendships from your work as a scientist. Nevertheless, I was determined to enjoy this final hurrah as a doctoral student while it lasted.

The field trips attached to botany conferences usually involve hiking, but are decidedly not for those who wish to move with purpose. Hiking with botanists, or "botanizing," as we call it, involves walking a few yards at a time with your eyes planted firmly on the ground, scanning for any vegetation that might be worth further inspection. When you find some, and you will, because most vegetation is interesting to botanists, you pull out your hand lens and get down for a closer look. If mosses, lichens, or small, herbaceous plants are your thing, you'll probably end up laid nearly flat out on the ground at some point, an acceptable position regardless of age or station. Particularly interesting finds will garner circles of bent-over bodies craning for a good look. Orchids are always a hit, as are mushrooms, despite not being plants, but each scientist has their pet plant group, from tiny liverworts up through gnarled old junipers and towering firs. Everyone is cheery, and we progress at a rate of tens of yards an hour this way until it gets dark or it's time for a snack or packed lunch. These are my people.

Helmcken Falls was gorgeous. A punchbowl-type waterfall with a wide, enclosed basin surrounded by forest, it has the fourth largest drop in Canada and looked perfectly boreal with its frame of pine and fir trees. The trails that had been selected for us near Helmcken Falls were unchallenging, to accommodate all ages and abilities of botanists, and the weather was perfect. The woods near the trail quickly turned into a smattering of crouching sun hats. But on this field trip, I saw something new: A woman was painting. She knelt with a small, landscape-oriented sketchbook and a tiny metal palette of watercolours, looking out at the falls. I'm not usually one to approach strangers and strike up a conversation, but this time I couldn't resist. Dressed comfortably for the field and looking to be in early middle age, with a kind face, she wore glasses and had orangey-red hair pulled back into a ponytail. Lyn Baldwin introduced herself as an ecologist at the host university in Kamloops. Over a rambling chat that continued on the old school bus that drove us back to Kamloops, I told her about illustrating my thesis, and she taught me about nature journaling. She would visit natural areas with her sketchbook and paints and capture both vistas and details of the flora and fauna in watercolour and ink, alongside written descriptions of what she'd seen and where. Beyond simple field sketches, I had never really considered taking art supplies out into nature to capture it firsthand and was taken with the idea. She must have sensed my enthusiasm as I perused the pages of her sketchbook, because she invited me back to her office at the university to look through her other sketchbooks. I jumped at the chance to try to learn how to do what she was doing.

She led me to a dimly lit room that seemed to be an undergraduate science lab. I soon understood why she needed the space. On the workbench she had set out *stacks upon stacks* of her filled sketchbooks. Years and years' worth of her artwork in hand bound books. She talked briefly about the sorts of places she sought out for her paintings, which gave me the impression she used these books only for nature drawing. She then told me she had meetings to attend, but that I was welcome to stay as long as I liked looking at the books. And then I was alone. I stayed for hours with this treasure trove, wanting to see every page. The books were set out more or less chronologically, and I watched her artwork evolve and mature. I watched her experiment with new styles and layouts and incorporate them into her practice. I itched to get out and try this myself when I got back to Montreal.

Though indeed the vast majority of the paintings were plants, wildlife, and natural landscapes, there would occasionally be a spread that showed a more human scene. A room or some people, the contents of her bag, laid out and sketched just for fun. The words on these pages read like a journal rather than a naturalist's log. They seemed randomly scattered throughout her books. I didn't dwell on them, because it felt like an intrusion; she'd invited me to look at her art, not read her diary. Maybe she'd forgotten those pages were in there. Maybe she didn't care. Either way, I tried to respect her privacy. That is, until I came across something that captured my attention in a way I couldn't ignore.

Over several entire page spreads in her neat, even handwriting, Lyn described the experience of carrying and giving birth to her daughter, Maggie. She wrote about the heavy and anticipatory final months of gestation, of wanting to shrug off the "inertia of late pregnancy." She meditated on whether she already loved this child, still weeks away from being born, and didn't know. She imagined the type of mother she wanted Maggie to have, and wondered how to be that person while still meeting her own needs and desires. All the things I envisioned myself asking and wondering in her place. "I am willing to allow my life to be subsumed beneath the immediacies of the upcoming birth," she wrote. *Would I be, if it were me?*

She recounted the feelings, emotional and physical, of squatting down to rest while on a hike and having her water break in "a moment of sustained disbelief." Of sensing her contractions begin, first gently, and then, later at the hospital, "with some teeth." She felt an air of unreality about what was happening to her, as though her body was making the transition from pregnancy to motherhood faster than her mind could keep pace. With imagery ranging from torrential water to a runaway train, she described the increasing intensity of her labour, peaking as she was finally ready to bear down, when the pain became a tidal wave she could ride to the shore with each push. Near the end, she recalled the Bene Gesserit Litany Against Fear from *Dune*: "I must not fear. Fear is the mind-killer . . . I will face my fear. I will permit it to pass over me and through me." An incantation used by powerful women to calm and focus their minds in frightening situations. And finally, the singular sense of relief as her daughter's body left her own.

"I feel shattered to the core, privileged beyond belief, and simply in awe," she wrote in the first days of her child's life as she watched the baby change before her eyes. "I never fail to be surprised at how infinitesimally small changes add up to a profoundly different being."

Lyn's words left me thunderstruck. Would I want to feel *shattered to the core?* I hunched over the book in the dim, silent lab, willing that no one would enter and interrupt me before I could fully digest what I'd read. I felt immediately ashamed, as though I'd looked at something I shouldn't have and might be scolded for it. Illicit knowledge. Her words were so visceral and honest. I was in a room alone with this recollection and I couldn't look away. It was as though I'd stumbled on the forbidden secrets of a group I wasn't a part of—mothers. The shame grew from having read the private words of someone I'd only just met, yes, but it was more than that. It ran deeper than that. I realized I'd never before spoken to anyone who had given birth about their experiences, had never even heard anyone describe it. How could I, a biologist so in love with the living world and its hidden details, have had such a glaring blind spot about a fundamental part of life? Why had I never even wondered?

I guess, like so many "female" things when you grow up motherless, no one was volunteering this information, and since I'd assumed I didn't want to be a mother myself, I'd never sought it out. To be suddenly confronted with it was to be confronted with a gap in my own awareness. But also, I guess I'd always figured that by the time a grown woman came to be having a baby, there were no more mysteries, no big surprises awaiting in that process. Adults just knew these things. But Lyn seemed as green as I was, so surprised and humbled by what was happening to her.

Still staring awestruck at the pages, I caught sight of a date and realized with a start that the baby whose birth I'd just witnessed was now nearly a teenager. I wondered how her mother spoke to her and thought of her now. Does the tenderness and magic remain through so many years? Were they still so close? Or does the shine get worn off by years of the usual skirmishes between parent and child? Had this baby grown into someone like Sarah—kind and easy to love?

As when I'd watched Pat's daughter, I felt my perspective shifting from child to parent. I stood in a moment, frozen in time, of seeing both youth and adulthood. The instant of stillness at the top of an arc, when I could look both forward and back. I imagined myself in Lyn's position, undergoing the monumental task of bringing a life into the world. That the body I'd heretofore thought so little of could do something so momentous. I asked myself if, as someone who wanted to experience life's miracles, I could really part with the gut-level distain I'd developed for the daily realities of motherhood. The small indignities of dirty diapers and sticky fingers and wailing tantrums. I didn't know.

ONCE IN A WHILE THROUGH the aeons of evolutionary time, a newly evolved feature turns out to be a significant improvement over the previous ones, at least for that specific time and place, and it proliferates enough to become the standard itself. If that new feature is *particularly* effective and leads to a big bump in fitness and reproductive success, it's what we call a "key innovation"[10] and can lead to bursts of new species as variations on the new and successful theme emerge. The ultimate in key innovations is flowers themselves.

We still don't know the specific series of evolutionary events that created flowers. What we do know is that at some point prior to the early Cretaceous period,[11] the naked ovules and separate male and female reproductive structures that were characteristic of seed plants up to that point[12] began to condense into a compact assembly of both organ types as well as some leafy, nonreproductive structures—what we now think of as petals and sepals. The ovules became enclosed in the protective structure of the ovary—the same tissue that forms the edible part of a fruit like an apple, and so true "fruit" was born. The piecemeal changes required to fully make this transition would have taken an enormous amount of time, but when it was complete, the plants that bore these early flowers had all the tools required to develop complex relationships with animals to achieve their pollination goals. In exchange for a bit of sugar they'd conjured out of thin air and sunlight, the plants could manipulate any number of different animals into carrying their pollen to an appropriate mate. These relationships would shape the further evolution of both the plants and the animals right up to present day.

The advent of flowers was so wildly successful that the flowering plants, known as angiosperms,[13] now make up over 90 percent of all plant species.[14] Their success spawned new species after new species, an ecological juggernaut that exploded across the globe and pushed the once-dominant ferns and gymnosperms to the margins, never to return to their former glory. The fossil record, our primary source of information about these events, shows that the Earth went from no flowers to flowers nearly everywhere with astounding speed between the beginning and the end of the Cretaceous period.[15] This apparent leap greatly troubled Charles Darwin, a proponent of the very gradual, incremental evolution produced by natural selection, because it appeared to have occurred so abruptly. He famously referred to the speed of the angiosperms' global takeover as an "abominable mystery."[16] What seems likely today is that there were a great many incremental, behind-the-scenes, preliminary adaptations—missing links, if you will—that took place before the first unequivocal flowering plant fossils, but were not caught in the notoriously incomplete fossil record,[17] producing an overnight success that was actually millions of years in the making.

The morphologies of flowers and their pollinators are sometimes so highly coadapted that the existence of one can imply the existence of the other. Famously, in 1862, Charles Darwin observed the star orchid *Angraecum sesquipedale,* with its strange, foot-long, tubelike nectar spur projecting out behind and below the flower, and predicted that there must be a moth, coevolved with the flower, that had a proboscis long enough to reach the end of the nectar spur, where the flower's sweet reward lies. He was mocked for proposing such an absurd thing existed. It took decades for that moth to be found, and indeed, it didn't happen in Darwin's lifetime, but it turned out that his prediction was correct.[18] The hawk moth, with its improbably long proboscis, was named *Xanthopan morganii praedicta*, Latin for "predicted moth" in recognition of Darwin's foresight. It's a powerful demonstration of the predictive capacity of a theory that can seem pretty abstract, since we usually can't see evolution happening in real time.

In the case of highly species-rich groups of plants such as orchids, you can often point to one or more major adaptations that contributed to the success of the group. Orchids have a unique flower in which the male and female parts are fused into a single structure, the column, and the pollen grains are bound up into big, waxy balls called pollinia. A modified petal referred to as a lip provides a landing spot for insects. All these adaptations allow members of the family to have very specialized relationships with their pollinators.

Some species have coevolved to attract a very specific pollinator, as in the case of the bee orchid *Ophrys apifera*. The flower mimics the body of a female *Eucera* bee, causing the male to attempt to mate with it, pollinating the flower in the process.[19] Such a high degree of specialization is risky, though, because the plant is dependent on a single insect species for outcrossing. In fact, this orchid is known to self-pollinate heavily,[20] which isn't a good long-term strategy; its morphology is presumably too highly geared toward one group of bees to be able to pivot to another, and it may now be on the slow road to extinction. This is a good illustration of how coevolved plant-pollinator relationships can be highly successful, but also very vulnerable should the relationship be disrupted, as they increasingly are in our current age of destructive human activity.

Like leaves becoming flowers or flowers becoming tendrils, myriad tiny, invisible changes can add up to a large, visible, sometimes astounding transformation. But if you go too far, adapt too profoundly, you may find yourself committed to a path that's no longer viable.

If you change *too* much, it's hard to go back.

AS THE SUMMER EBBED AWAY following my trips to New Orleans and British Columbia, I looked to the year ahead. My sheltered time of drawing plants and ignoring the world was ending. I'd spend the coming months analyzing all the data I'd collected over nearly five years and writing my thesis. In the spring, I'd defend my work before a roomful of my peers, and the project would be over. And then . . . For the first time in my life, there was no clear "and then" to lean against. Unless I could find a postdoc and a direction for myself, my progression from one tier of academia to the next was coming to an abrupt halt. My own moment of transformation, of tiny advances adding up to something big, needed to come soon.

Just behind these thoughts, my mind kept turning over images of Sarah with her family, and of Lyn's journal entries. Of mothers and daughters and a different sort of transformation.

Like studying natural history, motherhood was a way of knowing a part of existence through direct physical experience. Not through theory or intellect, but through touch and sight and other senses. It was the immediacy of knowing the world through our bodies. Gradual in its onset, but visceral and irreversible in the changes it wrought. Like the years of my life I'd given over to study, it demanded sacrifice in order to know. I could peek into another of life's precious secrets, but I could never come back. In a thought so dangerous and disruptive in its implications that I couldn't quite bring myself to probe it yet, I suspected my vision of my future might be starting to change.

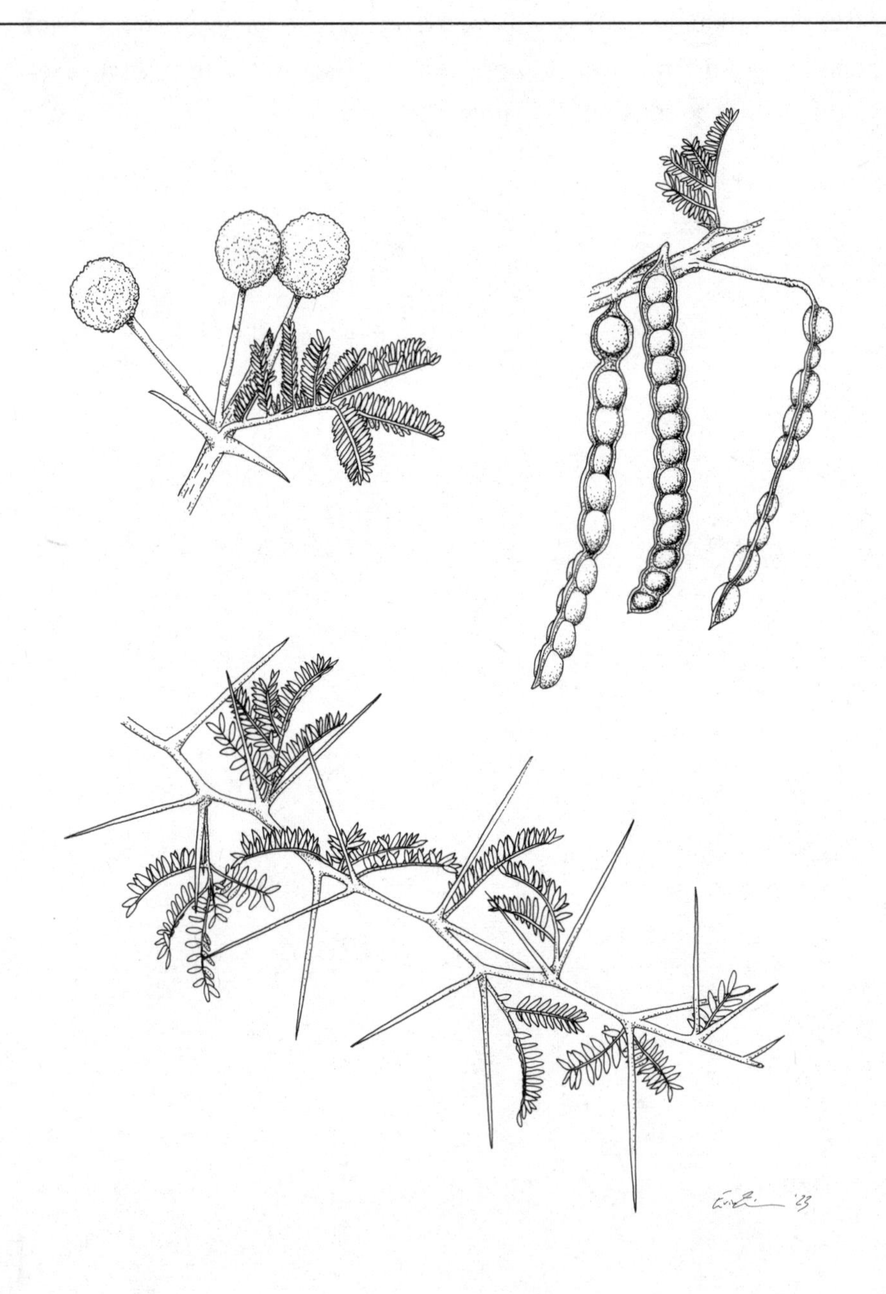

CHAPTER 6

UNCERTAINTY, IN SCIENCE AND LIFE

Given the extreme particularity of species and how little we know about them as a whole, taxonomy can justly be called the pioneering exploration of life on a little known planet.

—EDWARD O. WILSON

Sitting down to write a doctoral thesis is an intensive experience. It's just you and your research, alone together for months. It's solitary in a way that lab work, with its casual chats and group playlists, isn't. It's also your last chance to do your research justice, to be insightful . . . to show some hint of brilliance that might mark you for a bright future to some hiring committee down the road. For an obscure plant group like mine, it was possible that no one else would take any notice of it again for a long time, so I wanted to do right by my trees, to leave them well-classified and better understood and appreciated than they had been before. This thesis was my long goodbye.

For over four years, I'd given my whole attention to these plants, had come to understand them inside out—at the level of leaves and flowers and fruit, but also down to their very DNA. Yet oddly, I'd seen them only in death. I knew them better in this form: either dried and flattened, brown against the white of the herbarium sheet; or pale to the point of near translucence, adrift in a phial of chemicals that would never allow them to decay, but robbed them of any lifelike quality. I knew only the corpses and ghosts of these trees, but I knew them very, very well. Soon they'd become even less substantial to me, going from solid objects in my hands to only description and discussion on a page. Those mere words hid so much—the beauty of the living plants and their idiosyncratic solutions to life's challenges, the threats they faced, and their unimaginably long history. The more I read as I pushed through my thesis, hunting down the original species descriptions in decades-old publications, the more I realized that even the names themselves hid stories.

Take the genus *Uittienia*, with its single species, *Uittienia modesta*, named for its modest, unassuming appearance.[1] A tree that grows only in parts of Malaysia and Indonesia, *Uittienia* is very similar to members of the genus *Dialium*, and at times has been considered a part of that genus. However, unlike *Dialium*, which mostly has very few or no petals at all,[2] *Uittienia* has a full set of five petals.[3] The delicate little flowers sit in small sprays, likely pollinated by the wind. Hence the modest appearance—no pollinators to attract.

Uittienia was described and named as its own genus, but some taxonomists regarded it as a species of *Dialium* and renamed it *Dialium modestum*. It was a small matter of a single, obscure species, and as such, it had been left more or less unresolved and untouched for several decades. There were bigger fish to fry in legume taxonomy. But if the tree is indeed shown to be part of *Dialium*, its original name vanishes. This is a case where I think that matters.

Plants have their own stories, but very often, they have human stories attached to them as well. For example, the fern genus *Gaga*,[4] named for the musician Lady Gaga, or *Caloplaca obamae*, a lichen named for Barack

Obama. These are big, flashy examples. But sometimes there's a lesser-known story attached to a more obscure plant that deserves to be remembered. Like the story of Hendrik Uittien.[5] Born in the Netherlands in 1898, from a young age Uittien loved being out in nature, and particularly adored collecting and drying wild plants. He went on to work as a herbarium curator and lecturer at Utrecht University, as well as teaching at an agricultural school. His work centred around systematics and morphology, particularly of sedges, but also included studies of Dutch plant names and folklore. During the late 1920s and early 1930s, Uittien became an internationally respected botanist; his work was considered part of a revival of interest in Dutch nature happening in the early twentieth century. Uittien appreciated the more common plants like dandelions as much as the rarer ones, believing that nothing in nature is ugly.

When the Nazis invaded the Netherlands in the spring of 1940, he was disgusted by their ideas of racial purity. Though a gentle soul, Uittien engaged in small acts of resistance through the symbolism of flowers. In his yard, he planted nasturtiums, *Tropaeolum majus*, a symbol of patriotism due to their bright orange colour. He also wore a white carnation in his lapel—a symbol of the Dutch Prince Bernhard—on the prince's birthday, showing his loyalty to the royal family. Uittien was dismissed from his teaching position in 1941 after giving an anti-Nazi speech at the university. Following the dismissal, he became a deliverer of the illegal resistance newspaper *Trouw* and taught housewives struggling to feed their families under food rationing about local edible plants, all while still collaborating with fellow botanists on *Flora Neerlandica*, a flora of his country. In early 1944, Uittien, along with twenty-two other *Trouw* couriers, was arrested and eventually sent to the concentration camp Vught. Writing letters to his family from inside the camp, he downplayed the misery of the place, instead speaking hopefully and even describing a rare flower he'd seen there. He expected to be detained for only a brief time, given his "crime." His hopes for release weren't to be, though. On August 10, 1944, Uittien and the rest of the "23 of *Trouw*" were executed by the Nazis for their work with the newspaper. The genus *Uittienia* was named in his honour in 1948.[6]

Few outside the Dutch-speaking world have ever heard of Hendrik Uittien, and little has been written about him in English, but the genus named for him stands as a small memorial to his life.

THE TWIN SCIENCES OF TAXONOMY and systematics[7]—taxonomy names, describes, and classifies, while systematics arranges life in its family tree—have been works in progress for thousands of years. Initially, groupings were built around food and medicinal uses. The earliest records of such work come from Egypt and China in the form of pharmacopoeias and go as far back as 3000 B.C.E.[8] In the Western tradition, at least as far back as the Ancient Greeks, humans have tried to find and document useful, logical groupings and names for the plants that surrounded them. Theophrastus, working in the fourth century B.C.E. and often referred to as the "father of botany," considered various means of grouping plants, including their growth habit (trees, shrubs, sub-shrubs, herbs), lifespan, and whether they were cultivated by humans.[9] His *De Historia Plantarum* includes a classification of nearly five hundred plants and contains some genus names still in use today. Other Greek and Roman ancients, including Dioscorides, produced further works that added to this count. As with botanical illustration, the description and classification of living things largely ground to a halt during the Middle Ages. Herbals were produced but copied the information given in the works of the ancients.

With the Great Navigations of the fifteenth through seventeenth centuries, many more plants came to the attention of Europeans and needed to be organized into useful groups; there arose a need for a system that could accommodate larger numbers of species. The invention of herbaria around the same time aided in this task because it provided a concrete object to which a name could be attached, something that could be returned to for study as possible groupings were evaluated.

Classification of plants became more sophisticated with the introduction of "natural" systems of grouping—those based not on artificial criteria such as medicinal properties but on physical similarity. The first was proposed by Italian botanist Andrea Cesalpino[10] in the sixteenth century and established the plant families now known as the Leguminosae (beans) and the Asteraceae (daisies). Cesalpino's classification used only a very few physical characteristics to form groupings, however. His work was greatly improved upon by John Ray in the late seventeenth century. An English botanist, Ray understood that the best classification would come from using as many characteristics as possible. This allowed him to gain a detailed understanding of natural groups. He was the first to define monocots (one seed leaf) and dicots (two seed leaves), which still stand as fundamental groups in the plant kingdom. Most importantly, Ray proposed what is very nearly our modern species concept—that of interbreeding individuals whose shared characteristics could be seen in their offspring—at a time when that idea was still very blurry and inconsistent in scientific writings. Ray eventually classified over 18,000 plants into 125 different groups, and many of these are still intact as modern plant families. Ray put botany on a much more scientific path than it had been, all while working in the framework of natural theology; he sought to better understand God's handiwork. It would still be another century and a half before God began to be removed from the equation.

Throughout Cesalpino and Ray's time, plants were referred to using Latin polynomials—a string of Latin words that gave a quick and dirty description of what the plant looked like. For example, the tomato: *Solanum caule inermi herbaceo, foliis pinnatis incisis*, which translates to "Solanum with a smooth herbaceous stem and incised pinnate leaves."[11]

The names were unwieldy, however, and difficult to memorize. They were also not standardized from one region to the next, leading to a great deal of disarray and confusion when attempting to communicate about plant species. The situation wasn't remedied until the mid-eighteenth century, when a Swedish naturalist called Carl von Linné, Carolus Linnaeus in its Latin form, came along and proposed both a new way of classifying plants and a new way of naming them.

His naming convention you're likely familiar with. Linnaeus suggested what's called a binomial name, a two-part Latin name consisting of a genus and specific epithet (the "species" is actually the two words put together, not just the second one). While the binomial could be somewhat descriptive, its primary function was simply to signify and be easy to remember. Think *Aloe vera* or *Ginkgo biloba*, for example. Having come up with binomials, he then assigned more than six thousand of them to various plant species in a text called *Species Plantarum* in 1753. He must have assumed this would give him plant-naming supremacy in perpetuity, since he believed there to be only around 10,000 plant species in the world, as opposed to the 350,000 or so that we know of today. Binomial names were later extended to animals and other organisms as well, but plants were the first.

It's *less* likely that you've heard of Linnaeus's plant classification system, as it's now slipped into historical obscurity. Another artificial system—these types of classifications didn't disappear until acceptance of evolution ultimately rendered them indefensible—it didn't reflect actual relationships or even broad similarities between species, but was meant to be easy to use for both botanists and laypeople. And indeed it was. The "sexual system," as it was referred to (causing significant pearl-clutching among botanists of the time), was based on the number and arrangement of the sexual parts of the flower—the stamens and pistils. A person wanting to know what group a plant belonged to had only to count parts and perhaps note the lengths and arrangement of the stamens. Linnaeus knew it was an artificial system but believed that a true natural system was only possible once *all* plants had been discovered; he created his system to be simple and to stimulate public interest in botany in the meantime. Ultimately, the system fell out of use, but not before provoking one of the more amusing tidbits of eighteenth-century science history—an epic poem by one Erasmus Darwin.

Physician, evolutionist, free-thinker, and probable atheist before it was cool,[12] Erasmus Darwin, grandfather of Charles, also wrote poetry. Bad poetry, according to the likes of Samuel Taylor Coleridge and Lord Byron, his contemporaries in the late eighteenth- and early nineteenth-century arts scene.[13] His creations were science-inspired, and in 1789, he published *The Loves of the Plants*, first by itself and eventually as part of a larger work called *The Botanic Garden*. *The Loves of the Plants*, which attained fame in its time for its colourful language and vivid imagery, was a lengthy poem based on (and breathlessly admiring of) Linnaeus's classification system. In a series of verses touching on eighty-three distinct plant species, the poem dramatized the different arrangements of stamens and pistils, imagining them as men and women, sometimes in situations that were surprisingly racy for the time.

For the plants with one stamen and one pistil, Darwin started with a pretty vanilla picture of a virtuous husband and wife, but from there moved on to a couple sneaking around but betrayed by the appearance of an illegitimate child—"the green progeny betrays her loves"—and a woman awoken by an enamoured lover—"Rise, let us mark how bloom the awaken'd groves, And 'mid the banks of roses hide our loves"—among others. Among plants with multiple stamens and a single pistil, Darwin wrote of a lady attended by two squires and adored by two knights bent before her "fragrant altar," a frantic queen avenging rejected love by committing infanticide, a seductive harlot, and a beauty guarded by fond brothers.

Once he reached the classes with higher numbers of stamens *and* pistils, the old boy just couldn't seem to keep a handle on his imagination, describing "wanton beauties" in "gay undress," a "glittering throng" of "beaux and belles" and "a hundred virgins" joining "a hundred swains." In fact, the word "virgin" appears eleven times in the poem. I counted. It's little wonder the poem became so popular in such a buttoned-up era.

Unlike his contemporary Linnaeus, a pious believer in a vengeful God, Erasmus Darwin was a liberal, with two happy marriages that produced twelve children (though two did not survive beyond childhood) and relations in between those marriages that produced yet another two. An obituarist described him as never being able to "resist the charms of Venus." In fact, at the time of writing *The Loves of the Plants*, he was in love with a local married woman, Elizabeth Pole, who was later widowed and became his second wife. His colourful poetry might have been, in part, an attempt to woo her. Regardless, Darwin's *stated* aim in writing his scientific poetry was "to [e]nlist Imagination under the banner of Science,"[14] which seems to me a noble goal that excuses a certain amount of bad poetry.

Physical similarity between species was not fully understood to be due to common ancestry until Erasmus's famous grandson proposed evolution by natural selection. Only then did classifications become genealogies, as Charles put it.[15] Still, natural selection didn't change the *practice* of taxonomy all that much, since it didn't dictate any new techniques, it merely provided a new lens through which to understand the results of classifications. It helped us to understand *why* species were similar or dissimilar to varying degrees. As science historian Jim Endersby wrote in *Science*, "If Darwin was right, classification was much more than mere naming; it was uncovering the history of life on Earth."[16]

Taxonomy remained based in morphology until technological advances in the mid- to late twentieth century allowed evidence from cell biology and genetics to contribute as well. The introduction of DNA sequence data has been called "era splitting" for taxonomy[17] and has led to the discovery of many groups that hadn't been previously detected because they looked superficially similar, while causing others to be abandoned when it turned out that their apparent physical similarities were misleading.

Today, taxonomic determinations and classifications are primarily based on a combination of morphology and DNA sequence variation, though data on things like reproductive systems, flowering times, spatial distributions, as well as ecological and climatic niches can also inform taxonomists' decisions. Creating and abandoning names and groupings is a perennial activity in the taxonomic world—usually with good reason. New evidence may dictate breaking an established genus into multiple genera, for example, or moving a species to a different genus, which would change the first half of its Latin binomial name. These refinements are crucial because all research conducted and information gathered on those species will be built on those names and classifications. Conservation measures are also built on taxonomic knowledge, and not having fully grasped the species richness of a region can lead to species slipping through the cracks.

"Taxonomy is really the foundation for all of biology. If we're not using the right names for things, we're not communicating properly," says Pamela Soltis, a curator and distinguished professor in the Florida Museum of Natural History at the University of Florida, who also directs the university's Biodiversity Institute. Soltis's research focuses on the evolutionary history of flowering plants and their relationships to one another. She explains that for a lot of species, all we really know about them initially is what we can see on a herbarium specimen and its label, which may both be decades or even centuries old. So the plant may be classified with relatively little to go on, but as we learn more about its evolutionary history, distribution, or interactions with its environment, it sometimes becomes clear that the name needs to be updated.

"Continually revising and modifying our taxonomy based on our understanding of evolutionary history is really critical to having the names mean something. People in other areas of biology absolutely hate this aspect of taxonomy, and they resent systematists and taxonomists for changing the names on them all the time. But it's really critical that once we learn something new about a species and determine that the name that's been applied is not reflective of that new information, then we really need to change it, because otherwise we're doing a disservice. We're conveying information that's not correct."

There is, however, a certain amount of wiggle room for personal philosophy among taxonomists because biological species as a concept aren't as cut-and-dry as we're taught in school—there is room for interpretation, and the lines between species are frequently not entirely distinct and obvious. Some taxonomists, the lumpers, tend to focus on similarities and lump more organisms together into a single group. Others, the splitters, tend to focus on differences and divide groups more finely. Splitters look at nature and see a greater number of different species than do lumpers. The lumpers see a greater amount of variation—subspecies and varieties—*within* a single species. There isn't really an objective way to say that one view is better than the other, and each group tends to stubbornly defend their philosophy. Charles Darwin considered the perennial differences of opinion between lumpers and splitters to be evidence for evolution as a process.[18] If organisms were always changing and evolving as he predicted, the boundaries between them *should* be blurred. If, on the other hand, they had been divinely created in their current form according to the teachings of the church, one would expect the boundaries between them to be quite distinct.

Differences in how various taxonomists see and define species has led to uncertainty in estimates of the number of species that are out there. Among plants, the number of known, officially described species varies anywhere between 350,000[19] and 374,000,[20] depending on how those species are defined. If we broaden our view, the margin of uncertainty gets much larger: the total number of *named* species on Earth is somewhere in the vicinity of 1.2[21] to 1.5 million[22]—an uncertainty equal in size to the entire plant kingdom. And if we then consider those nonmicrobial species that are yet unknown but projected to exist, the number tops out at anywhere between 2 million and 8 million.[23] Finally, if we wish to include all the microbe species predicted to inhabit this teeming Earth, that adds an almost unthinkable one *trillion*[24] further taxa to deal with, and these last get very tricky to define indeed. It's a big, complicated, confusing world, and the work of taxonomists sits right at the centre of our efforts to understand it.

Species concepts play heavily into the question of whether taxonomists will ever succeed in their ultimate goal of naming all of Earth's species, Soltis explains. "What *is* a species? Is the species a biological entity, or is it a human construct?

"It's our ability to put a boundary and apply a name to something. But what we call a species in one group, in terms of how much it varies from another group, might be very different from what's used in other things." Soltis uses the example of oak trees, which are known to hybridize frequently between species. Being aware of this, oak taxonomists seeing hybrid trees aren't concerned that it's an indication species are wrongly defined. In other groups, however, where hybridization is rare, if it did occur, taxonomists would need to ask if the two parent species are not, in fact, two different species. "So there's a lot that goes into determining where to recognize those boundaries, and as we get more and more detailed data about organisms, where do we draw the lines? How do we recognize all of this diversity? Because the goal is to be able to communicate about what is in nature.

"Throughout my whole career, they've said, 'Something is going to solve this problem.' It was flavonoid chemistry, then it was allozymes, then it was chloroplast DNA, then it was blah, blah, blah. And now we've just realized that . . . everything's too complicated." Soltis says she now realizes that there isn't going to be a single, silver bullet approach that "solves" the problems of taxonomy. It will always need a multifaceted, holistic approach, but she believes that's what keeps the work stimulating. "As a systematist, you need to know a lot of stuff. You need to know some ecology. You need to know some genetics. You need to know morphology. You need to know all of these different things, and it makes your work extremely interesting, but it also means that it requires a lot of training for people to be able to combine these sorts of knowledge."

It's easy to imagine that something so arcane as the bestowing of Latin names and grouping of species doesn't have any real importance or impact on life outside the ivory tower, but for an example of how scientific debates can bleed into the "real" world, look no further than what became the biggest and most media-documented dispute in modern taxonomic history. Acacias are trees of the legume family, often shrubby, sometimes spiny, with recognizable shapes and fluffy bursts of little yellow pom-pom flowers. Iconic to the landscapes of both Australia and Africa, "the wattles," as they're known in Australia, are a dominant species over much of the continent and are the national floral emblem; in southern and eastern Africa, meanwhile, they dot the grasslands amid such quintessentially African creatures as giraffes and lions. The image of a flat-topped acacia tree silhouetted in front of a fiery sunset is emblematic of the African savanna.

Botanists had known since the 1980s that the large genus *Acacia* should be split into several smaller ones based on differences in their appearance, and the emergence of supporting DNA evidence in the early twenty-first century made the issue impossible to ignore. But the split had consequences. Under the official botanical naming rules, only one group could retain the genus name *Acacia*. Normally, that would automatically be the African group, because that group contained the original type species—*Acacia nilotica*—used to define the genus. This species, chosen at the time the genus was established in 1754,[25] acts as an anchor for the name, which goes with whichever group that *A. nilotica* belongs to; the genus was *based* on an African species. According to obscure naming rules, all Australian species would have to be renamed under the genus name *Racosperma*.

However, a special provision in those rules allows for the type species to be changed in cases where *not* doing so would cause difficulty or confusion due to a large number of species having to be renamed. Since the Australian group consisted of around one thousand species[26]—fully 6 percent of all plant species in Australia—while the primarily African (and to a lesser extent, Central and South American) group consisted of only a few hundred species, a proposal was made in 2003 to use that provision. Thus began what came to be known as the "Wattle Wars," unprecedented in modern taxonomy for both the number of species it affected and in the degree of public attention it captured. Both groups, Australian and African, faced a culturally significant, iconic name being stripped from their landscape, and people were *incensed*. Because names matter, and everyone knew it.

Australian arguments were based mainly on numbers and the economic importance of their acacias in forestry, as well as on its prominence in the landscape and national symbolism. Also, people just hated the name *Racosperma*. African arguments included the fact that, though fewer species would be affected if African acacias changed, far more *people* would be affected by the changes, and those people lived in countries less able to bear the associated administrative costs of the name change than wealthy Australia. And anyway, they pointed out, the very word "acacia" refers to the sharpness of the thorns on their trees, which, by and large, Australian wattles don't even have.

Much ink was spilled both in the popular press and in scientific journals as both sides made their case. The matter eventually fell to a vote at the 2005 International Botanical Congress in Vienna, where the Australian side ultimately won by a narrow margin, but not before the debate got quite hostile. Some tried to bring the matter up again for annulment at the next botanical congress in 2011, which, to their probable dismay, was held in Melbourne, and the move met with even higher approval once cooler heads had prevailed. Still, there were those who protested by refusing to use the new names for years, and as I write these words, there are botanists who are still angry about what happened.

The point is that names are much more than mere labels we slap on organisms to help us keep them all straight. They aren't neutral. Names carry emotional, cultural, and historical weight. Even the supposedly unbiased scientists who preside over them get caught up in their own feelings and attachments. As the authors of one chronicle of the dispute put it, "Taxonomists, it seems, are just as emotional as nonscientists when it comes to naming plants."[27]

But the bigger threat isn't too much sentiment over our classification of life, it's too little.

The most astounding thing about taxonomy as a science is the yawning gap between how crucial it is to the enterprise of human knowledge and how valued it is in practice. Notwithstanding its dismissal, along with the dismissal of other branches of natural history, as being a vestige of a time gone by, as we've seen, there are many, *many* species yet to be named and classified if we are to truly know our home. Beyond their freestanding value as living things, these are, to take a very human-centric view, where our future medicines, foods, and scientific innovations will come from. In order to describe, understand, use, and protect an organism, it must have a name.

Like other disciplines of natural history, taxonomy takes a long time to learn. The degree of knowledge required to know a group well enough to understand the subtle differences between species and whether you're seeing something new takes many years of study to acquire. And the fundamental, diagnostic work taxonomists do lacks the speed of modern science. The field has been struggling to keep up. Some have suggested that a more high-tech approach is needed, but the taxonomists themselves tend to point directly at the funding required to maintain adequate numbers of experts. In a 2021 editorial in the *Zoological Journal of the Linnaean Society* entitled "The taxonomic impediment: a shortage of taxonomists, not the lack of technical approaches," ninety-three authors signed on to make the point. "Many of the problems of taxonomy stem from the dismissive attitude of the scientific community toward this scientific discipline, its unwillingness to appropriately support taxonomic work, a concomitant elimination of academic positions and a growing suspicion towards fieldwork and specimen collection," they wrote, noting that "most great finds in science, from gravity to evolution, stem from description of observations."[28]

Soltis describes the changes she's observed during her career: "Departments of biology used to have a whole group [of taxonomists] . . . there might be an algal specialist, a bryologist [moss and liverwort scientist], a pteridologist [fern scientist], multiple angiosperm people, maybe even a gymnosperm specialist, along with their specialists in limnology, herpetology, ornithology, *et cetera*. Generally speaking, today, a department may have a couple of taxonomists at most, and they probably aren't focussed entirely on taxonomy for their research, which means that there's less and less time available for doing taxonomic work."

Damon Little, curator of bioinformatics at the New York Botanical Garden, describes the situation in similar terms. "It seems to me that botany generally, and natural history work specifically, has been slowly hollowed out in the sense that the funding sources have decreased and the job demands have increased. We've ended up with a lot of people [who] are experts in particular groups of [organisms] that don't actually have a lot of time to work on them." If you were to look at a systematist a few decades ago, he tells me, most of their working hours would be devoted to the plant or animals they had expertise in. But over time, a lot of additional tasks have been added to their jobs, and the amount of time they can spend looking at specimens has gone way down. This drop in taxonomic work has been compounded by a plummeting number of funded positions for those experts. "So we're losing expertise in all sorts of ways," he says.

Soltis agrees. "We're losing, we continue to lose, and we have been losing—*for decades*—taxonomic expertise. We're losing this expertise, and for lots of reasons—partly because people think that taxonomy, because it's been around for so long, is old-fashioned and not important. We're not doing a good job of communicating the opposite."

A major structural stumbling block to the preservation of taxonomy as a science is the way academia judges taxonomic publications.[29] A relatively small number of journals publish taxonomic descriptions—those that formalize a new species, for example—and these have relatively low impact factor, meaning that, in effect, publishing in them is less prestigious, and less useful when it comes time to be evaluated for tenure or research funding. Impact factor, a metric based upon citations during a moving window of the past few years, was originally intended to aid librarians in choosing which journals would be the most useful to subscribe to with their limited funds but has warped into a number used to justify decisions made about a researcher's career advancement.

The low impact factors are due to the fact that taxonomic descriptions are rarely cited by other papers. Unlike other research findings, these descriptions need not be cited when they are used in work that builds upon them. What's more, there's no consideration made for the fact that, while a "normal" piece of published research may only remain current and relevant for a few years, or even a few months, taxonomic descriptions can remain relevant for centuries. Therefore, despite being the absolute foundation that every other piece of research rests upon, they are rarely cited and therefore appear, in terms of their metrics, to not be of great value to the scientific community. A researcher could have many publications to their name, but if they're all taxonomic, that person may not be looked highly upon by hiring and tenure committees. It's a strong disincentive to do this research.

AFTER MONTHS SPENT TAPPING AWAY at the keyboard, constantly referring back to one article or graph or data point after another in an effort to get it all right, there came a day when everything had been analyzed, written, and illustrated, and I had to admit I was done. That day arrived in late winter, and I walked bleary-eyed and tired onto the balcony of my apartment, breathing out clouds around me. It was frigid, but bright and still, and I could see broad swathes of ice still lining the St. Lawrence on the far side of the tree-dotted park across the street. By the time those trees flowered, I'd be gone. When you start a big project, you have a mental image of how perfect it will be, how flawless and complete. But this being real life, and the perpetually inexact science of biology, no less, the reality was something different. It was my best effort in a world of limited samples, incomplete data, and fatigue. It reflected the truth of nature through a lens of imperfect human effort, as all science does. I was disappointed I hadn't been better, hadn't somehow made it as perfect as I imagined it could be.

Science is so much messier than people imagine it to be. Results are presented in clean charts and graphs showing a black-and-white picture that looks so certain, as though all the pieces had simply been sitting and waiting to be properly fitted together. But the pieces are never all there, and even among the ones that are, some are broken, and some have had their distinct shapes worn down over time, making them harder to place. Sometimes you sit and stare at the best version of the puzzle you're capable of making and still can't clearly see what it's meant to be. So you do your best to explain what you can make out until someone comes along who can do better.

My evolutionary tree—what scientists call a phylogeny—of the Dialiinae was the most complete picture of the group to date. That is to say, it included more species and genera than any previous attempt had managed, which allowed us a clearer look at how the individual species evolved and are related. It also allowed me to speculate on what forces might have caused them to look the way they do. Take, for example the genus *Labichea*, a group of about fourteen shrubs native to Queensland, Australia. My phylogeny allowed us to see that the earlier-evolving species of the genus have compound leaves with broad leaflets, sort of like walnut trees, but smaller and more oval. From this, there is an evolutionary progression to species with fewer leaflets, becoming long and thin, covered with hairs to keep the parched wind from pulling water away from the surface. They became tougher and more leathery to the touch. In one species, the leaflets have narrowed and hardened and curled up on themselves so much that they are, for all intents and purposes, *needles* now. Lining up these species next to one another is like watching a plant evolve and adapt to the hot, dry Australian outback in real time. It shows what plants do to cope with these types of environments. It was so exciting to see the picture take shape.

Less exciting was what my research had shown me about *Uittienia*. Though it was one of the trees for which I could never get a good DNA sample, I had painstakingly studied its morphology and included that data. My analyses placed it squarely within the genus *Dialium*, meaning the data didn't support *Uittienia* as a separate genus. The taxonomic work of defining genera wasn't a part of my project, but when someone eventually comes to revise those groups, mine may be the research they'll point to when they say the name *Uittienia* should be discarded. It saddened me to have found that that was the case. But we can't reject our findings, even if we wish they were different.

It was never really realistic of me to expect that my analyses would give a perfect, clear result that just came right out and said, "This is exactly how these things are all related to one another." It was a difficult group to work with; I knew this. But I also wasn't ready for the level of ambiguity I was ultimately confronted with. I was going to have to walk away from the project with a lot of the questions I'd been chasing still unanswered. There were entire genera for which I couldn't say with any certainty how they fit into the evolutionary story of the Dialiinae. Scientific results are there to be carried on and retested when better methods become available. In the future, it may become possible to get good-quality sequences from the old, degraded samples I was working with. New collections may even become available.

It's always hard to look back knowing you missed an innovation that could have changed the outcome of your work—to realize that you just arrived a little too early to the problem. Before my PhD work, I did a master's degree working with the fungi that live symbiotically in plant roots called mycorrhizal fungi. They are extremely difficult to both culture and to meaningfully DNA sequence, and I spent several years attempting to sequence certain individual genes using the sequencing methods that have been in use since the mid-1980s. Ultimately, my results were ambiguous and limited by what I was able to do on a technical level. Within two years of my finishing the project, a revolution occurred in sequencing technology, bringing us what is now known as "next generation sequencing," which allows massive amounts of sequence data to be accrued at a tiny fraction of the cost of traditional sequencing. To have had that technology available to me at the time of my fungal work would have drastically changed the outcome of the project. But it wasn't to be, and all I could do was hope someone else took up the question after me.

It would be both encouraging and bittersweet to see something similar happen for the evolutionary history of the *Dialiinae* in the future. And it's likely that it will. Systematics is being revolutionized by ever-improving sequencing technology, and our understanding of how DNA evolves is always being refined. Taxonomy, meanwhile, is looking to computers to find its revolution.

CHAPTER 7

ADAPTATION

The urban landscape is not the native habitat for mosses or for humans and yet both, adaptable and stress tolerant, have made a home there among the urban cliffs.

—ROBIN WALL KIMMERER, *Gathering Moss: A Natural and Cultural History of Mosses*

Many plants have great difficulty in surviving these extreme conditions. This is evidenced by withered and dying cacti on dry ledges. It is interesting to note that the same species may also languish among boulders on deltas and sandy shores. Here is a case where drought vies with flood waters in exterminating plants struggling for existence in a trying situation.

—ELZADA U. CLOVER and LOIS JOTTER, *Cacti of the Canyon of the Colorado River and Tributaries*[1]

With fewer and fewer trained taxonomists available to assess collections and determine which may constitute new species, much less identify those that are of already-known species, plant taxonomy is turning to machine learning to take a cutting-edge approach to an old practice. As we've seen, a huge number of unidentified specimens sit in herbarium back rooms, sometimes waiting decades to be examined, and many species new to science are projected to be among their ranks—likely more than half of all undescribed plant species, according to a study conducted in 2010.[2] In the absence of adequate numbers of trained taxonomists, computers may be able to help make the backlog smaller and more easily targeted, enough so that the taxonomists we have available can make real headway.

If a computer can "look" at an unidentified specimen and group it by similarity with a genus, if not a species, with high accuracy, then the work needing to be done by human minds decreases enormously. Taxonomists are left to focus on specimens that cannot be identified with high confidence, and those will have already been narrowed down enough to be more easily targeted to the appropriate expert.

Damon Little is working with colleagues to create a machine learning–based program, iCurate, to do just this for the ninety thousand or so completely unidentified herbarium specimens[3] waiting at New York Botanical Garden. It's not *hard* work to do by hand, he tells me—he finds it fun; the problem is how long it takes.

"You might spend all day on one specimen or you might get through a couple hundred, depending on how easy or hard they are," Little says. "This tool, hopefully, will make it so that a lot of that work can be done [automatically], and the specimens given to the right taxonomic expert, labelled as 'we know it's this genus, but we don't know what species,' or something like that."

So how is this accomplished? Using a machine learning algorithm, a computer can be trained on a large data set of preidentified reference specimens. The algorithm can then apply that training to make "educated guesses" when shown new specimens that fall within the same groups it's been trained on. It's no simple task, however, because the visual differences between species can be very subtle. "These data sets have a lot of very similar features, which means that the machine learning algorithms have to learn to distinguish among things that are pretty similar," Little says. "It's not so hard to, say, tell the difference between a boat and a cat. I mean, it's hard enough with a machine learning algorithm, but it's not the most challenging thing. Whereas, say, distinguishing between five different species of moss that all basically look the same unless you look under a microscope . . . now that's a challenge."

Little and his colleagues have run annual programming competitions for several years, providing the curated data sets and challenging teams to create the algorithm able to most accurately identify unlabelled specimens. The results have been very promising. "The scores that we're getting are in the mid- to upper eighties out of a hundred. So pretty good. Not the kind of thing where you would want to rely on that identification for whether you decide to eat a [potentially poisonous] plant or not." He laughs. "But definitely not too bad. And we're slowly trying to figure out ways to make it better, partly by teaching the neural networks better, but also by trying to give the competitors more related data sets that they can use in the training." The data sets being used for training, though they run into the millions of specimens,[4] cover only a restricted geographical area and number of species. Broader data sets become unmanageably large, and good training images just aren't available for many groups of plants. "It's going to be a very long time before we have something that's worldwide," Little admits.

The algorithms the competitors are training are designed only to match unidentified specimens to one of the groups of plants they've been trained on. What if the specimen is actually new to science? "We have this difficult problem in machine learning, which is that, for example, the machine learning model is taught to distinguish between cats and dogs. If you then show it a picture of a bear, it's either going to say 'that's a cat' or 'that's a dog.' It can't say 'I don't know.'" This quirk makes picking out new species a special challenge requiring its own approach. To address this, Little is working on another program, which he calls Novum, though he says it's still a few years out. There are several approaches he could take, including having the computer output probabilities with its guesses or even generate simplified images of the main diagnostic features of the specimen so that novel features are easily pinpointed. Either approach will take time that Little struggles to find. "The technology exists," he says. But he adds, "Everything takes longer than you think."

Another avenue for making the most of existing taxonomic expertise has been the targeted training of new researchers. A short-lived but successful program run by the U.S. National Science Foundation, PEET, short for Partnerships for Enhancing Expertise in Taxonomy, sought to assist in the direct transfer of expertise between generations of scientists, Pam Soltis explains. "The idea was that there would be a partnership between someone who was maybe approaching retirement age, working on a poorly known group of organisms, paired with a young, or either early or midcareer person, who was interested in that same group of organisms, so that there could be some knowledge transfer before that person retired. And that was a very, very popular program. It didn't have a huge amount of funding, but it was definitely a popular program. And it helped."

Indeed, the PEET program ran from the mid-1990s through the early 2000s and funded projects to better the taxonomic knowledge of dozens of understudied plant, animal, fungus, and microbe groups.[5] Over two hundred trainees were funded, and each project produced numerous publications, totalling in the hundreds. Participants contributed heavily to Genbank, the publicly available repository of genetic sequence data, and nearly all the projects involved collecting trips to gather new specimens. Many used the funding to host collaborative visits by foreign scientists or taxonomic conferences, pay for imaging or databasing of type specimens not easily available to researchers, and train parataxonomists—local experts working on the ground in biodiverse countries. A special meeting of PEET recipients even helped develop new curricula and training resources for teaching taxonomy.

A similar program today would help stem a loss of expertise that has largely continued since the previous program ended, retaining some of the accumulated expertise of the many taxonomists aging out of active research. But real change requires an ongoing commitment to taxonomic research. Neither limited, stand-alone programs such as PEET nor ingenious tricks of machine learning will save the field from broad institutional and public indifference. "It's more of a stopgap," Little says of his efforts. "Ultimately the problem, the gap in taxonomic expertise, is not a scientific one. It's a political one."

Even Little realizes that solving the specimen backlog using computers could be a Pyrrhic victory. He knows that the technology, if it works well, can just as easily be used to justify cutting back the training and retention of taxonomists even further. "I think machine learning generally can be used as an excuse to degrade human knowledge and training, not just in systematics, in anything almost. And I think in systematics it's particularly acute," he says.

As my conversation with Little drew to a close, he made the point of emphasizing to me what a *human* endeavour taxonomy is, despite the entry of artificial intelligence into its midst. "This work is possible because of all of the people that did the work before, in terms of cataloging plants, describing species, putting labels on specimens, collecting specimens, digitizing specimens. It's this massive amount of labour going back hundreds of years that has led us to the ability to do this today. We often don't acknowledge that, but I think it's actually very important to emphasize because this is the continuous evolution of human knowledge rather than something newly created. A lot of stories about machine learning talk about it as being this wonderful new thing that just sort of emerged spontaneously, rather than coming from hundreds of years of work. And I think that's kind of important to remember. We as a discipline have been working on this for almost three hundred years."

YOU SPEND A LOT OF time imagining how your PhD will end. At most doctoral defences, after the presentation has been delivered and the questions all asked and answered, once the committee has deliberated and decided that the candidate has made a meaningful enough contribution to their field, there is a moment when the defender is called to the front of the room by their advisor and, in front of their assembled peers, told solemnly, "Congratulations . . . Dr. [LastName]," to a big round of applause. And it's the first time you ever hear yourself addressed that way. Often the person is so relieved and happy, they cry. At the same time, in this age of egalitarian informality, you're supposed to act like it's no big deal, like the title means nothing, but I'll admit it here—I dreamed about that moment for years. I imagined it in all the adoring detail that some people imagine their wedding day; how smartly dressed I'd be and how everyone would be smiling at me; how I'd beam back at them and feel like I'd finally won out against my own bottomless need to prove myself. How I'd tear up with happiness. With so much still uncertain about my career and where I would end up, it was the thing I held on to when nothing was working and it felt like I'd never finish.

My defence took place on a sunny afternoon in mid-April in the lecture hall of the Biodiversity Centre. It had been publicized for weeks ahead of the date, so plenty of people from the Centre turned up, including lots of friendly faces. I was so nervous I felt sick. Weeks of intense preparation preceded by months of writing my thesis and years of learning my field suddenly didn't matter when faced with the possibility that I'd be asked a question I didn't know the answer to. This is expected, of course. It's the committee's job, in part, to test the limits of your knowledge, and you will therefore likely run up against some questions you can't answer. At that point, your job is to admit you don't know and use your expertise to speculate gracefully. But knowing that doesn't make it any less terrifying.

I'd spent the preceding week getting more nervous by the day, and the morning of the defence itself double-checking my PowerPoint and willing myself not to throw up. When the proceedings finally began, the committee members and I were introduced. Dressed in my most academic-looking tweed slacks and fitted vest, I walked up to the front of the room holding a heavily bookmarked hard copy of my thesis, trying to control the shaking in my hands.

As with most things, the anticipation was worse than the reality. After a few minutes of speaking, I'd fallen into a rhythm and the quiver went out of my voice as my throat relaxed. I realized I could make the most of my last chance to talk about a subject I loved so much. By the time I finished, I was actually having fun, and my smile was genuine. As I wrapped up my talk and braced myself for the questioning to begin, my nerves resurfaced again, but only briefly. Defence committees are usually nice enough to throw you a few softballs to start because they know you're terrified. By the time they hit me with one I couldn't answer, I had enough calm headspace to slow down, take a deep breath, and speculate. Even this seemed almost fun in the moment. As high-pressure as a doctoral defence is, you've spent years at that point only being able to discuss your research with a handful of people in your field, and getting to take the time and play with those ideas openly in front of an interested audience is a rare privilege.

The committee eventually wrapped up their questioning. A few general questions from the audience later and it was time for the public and me to exit the room so the committee could decide my fate. Though it was clear by that point that it was *highly* unlikely that I'd failed, the ten minutes or so of milling around out in the foyer with the audience I'd just been speaking to was tense. My moment was coming, and I could feel my heart racing in anticipation. We shuffled back in, a procession headed for the altar.

"Would the candidate please come to the front of the room?" one of the committee members asked. My knees felt weak, and I struggled to walk naturally as I made my way to the front. I couldn't quite meet the audience's eyes now that I was just standing there with nothing to say, so I looked at the floor while I waited for my result. The head of the committee, a kind-faced ecologist in his fifties, stood up holding a sheet of paper, as though the verdict were long enough that he needed to read it. It was finally happening.

"You'll be pleased to know that the committee was very happy with both your grasp of the subject matter and your contribution to the field, and has voted to pass you. Well done!" he said, smiling.

Applause. A standing ovation.

Wait, what? Is that it?

And it was. Time carried on. I smiled dumbly in a mild state of shock while the moment I'd waited years for passed me by in so many unspoken words. Friends and colleagues came to congratulate me, photos were taken for posterity, and I moved dry-eyed through all of it with part of my mind still stuck on the scene that didn't happen, like a sweater caught on a nail. Well-wishers paused in the foyer for drinks, including the gin and tonic thoughtfully provided by Anne, who knew I was the odd one out in that French culture of wine drinkers.

Then everyone dispersed, back to their labs and their experiments, and after a quick but jovial early dinner out with my committee, the day was over.

The end of my time at the institute and in Montreal came not with a bang, but with a tired sigh.

If Eric's work meant being stuck somewhere and trying to find solutions and build a career in that place, southern Ontario was as good a place as any. I missed home. After more than a decade away and having rounded the corner into my thirties, the nomadic life had lost its appeal. There were plenty of universities and government research facilities in the region, and I could keep trying until I got hired as a postdoc in one of them. I wanted the constant planning and speculation to stop; I'd been running scenarios in my head for what I could do for literal years, trying to solve this problem, and I didn't want to do it anymore. If I had to brute-force a solution out of my birthplace, I would. I would make it happen, and that was that.

I imagined all the people back home who would soon be asking me what I planned to do with my fancy degree (where I'm from, people will always preface the word "degree" with the word "fancy"). I feared the amount of energy it would take to look as though I *had* a plan.

ONE OF MY CHORES FROM age eleven or so onward was to clean out the horses' stalls and re-bed them with fresh straw. Done without stopping, it was probably the work of an hour or less, but I always got bored or tired and took breaks that drew the task out considerably. One day I was lazing about in the haymow as the sunlight filtered in through the dusty windows of the old wooden barn, wasting time because I was sick of shovelling manure. I began picking apart the pieces of hay, cut from the grass of a local field. My boredom gave me the chance to notice the little things . . . like the way the grasses formed hollow tubes that were blocked here and there by hard bits that I couldn't force my nail through. The way the stalks came to a head full of what looked like bits of tiny flowers but had no discernible petals. As far as I could pick them apart and peer closely with my naked eye, there was more to see. It made me wonder if everything had so much hidden detail if you only took the time to look.

The work of an agricultural community has room for everyone, at every age. Kids can get full-time summer jobs working in the fields as young as twelve, and many do, because the idea of earning real money when you're twelve is irresistible. A popular—though physically punishing—choice, and the one I jumped on board with, is corn detasseling. It was an easy job to get, and if you stuck it out, you could make a few thousand dollars in two or three weeks. A lavish sum for a preteen. I was up at five in the morning through July and early August to pack a lunch and make it to the school bus waiting in a nearby village by six. My father got up and cooked eggs and sausage for me each morning as I got ready. Even as a kid I was touched by the kindness of this. There's no love like pulling yourself out of bed in the dark every single day because you want your child to have a hot breakfast. The bus would drive us, half asleep, out to one of the enormous seed cornfields of Ontario's southern tip and spill us out, blinking, into the sunrise . . . the children of the corn.

The idea of detasseling is to produce corn hybrids. To do that, the male flowers, or tassel, need to be pulled off of all the "mother" plants to prevent them from self-fertilizing the female flowers, the silks, further down the stalk. We are emasculating corn. The motion involves reaching above your head, grabbing a sheath of leaves still enfolding the male flowers, and yanking it hard upward. It is likely you have not used this particular set of muscles in this way before, and it will hurt at the end of the day, at least to begin with. Muscle aches are only one of many resulting complaints. There are also blisters and cuts all over from the sharp leaves that press in around you as you walk through the close rows, there are sunburns and insect bites and, if you're not careful, sunstroke. You will walk twenty-five kilometres a day doing this, seven days a week, until it is done, because, as the farmers love to tell you, "crops don't stop growing on the weekend."

The work is brutally hard but strewn with sublime moments. First thing in the morning, it's cool, and the fields are shrouded in sunlit mist over the rolling hills, like a dream of summer. Each leaf anoints you with its bit of dew as you walk by, a bath in an emerald sea, and you're quickly soaking wet. It's chilly at first, but as the day heats up, that early morning soak will keep you cool for a little bit longer as the sun burns the dew off workers and leaves alike.

By ten o'clock it is hot, and growing more oppressively so by the moment. The fierce light off the leaves seems to emanate a deep green so intense you feel like it will absorb you. The cornrows are high and scratchy and beautiful and endless, turning to a wavering mirage of pale blue sky in the distance. This monoculture of vision occupies your sight nearly all day every day, so that when you close your eyes to sleep, you see only ghost images of cornstalks. The other detasselers drift away, and you are left alone with your thoughts in near silence, with only the soft, rhythmic popping of tassels being pulled to lull you into a meditative state. You let yourself go and seep into the greenness, just another part of the field.

I went back every summer for three years and bought my first laptop with the money I earned detasseling. Eventually, the allure of a "real" job at the grocery store with friends from town pulled me out of my annual corn communion, but the joy of being utterly immersed in plants never went away.

And here I was, nearly twenty years later, still trying to stay immersed.

Eric and I had planned our wedding for only a month after our move back to Ontario. There was so much to do to get ready for it that our boxes sat unopened in the farmhouse we rented until well after the honeymoon. Enthusiasm for getting everything ready for the big day lightened my mood and created a space where, for a little while and for the first time in memory, I could think only about my home life. I couldn't believe the lightness of it, like I'd jumped off a cliff and floated instead of falling. Being back among the cornfields and gravel roads gave me a peace I hadn't felt in a long time, and a tight, anxious little spot in my mind devoted to trying to cope with the sensory onslaught of the city could finally rest.

We were married in my father's barn on the summer solstice. It was a gorgeous, warm, sunny day. Eric and I had been working on the barn for a year, doing a bit of cleaning and repair on each trip home to visit, transforming a musty, abandoned old haymow with a rotted floor into an airy wooden cathedral where sunlight streamed through the gaps between the barn boards high above us and we could finish the night dancing under strings of fairy lights. Nearly my whole family was there, though almost the entire "friend" component came from Montreal; I had few local friends left after being gone for so long. For me, the day was loaded not just with all the normal symbolism of a wedding, but with the feeling that this was my reintroduction back into my natural habitat. I'd missed home so much for so long that the place had taken on a nearly mythical quality.

It would be months before I started to realize just how much everything had moved on in my absence. Off chasing my dream of becoming a scientist, I'd missed out on thirteen years of potlucks, barbecues, and cups of coffee around the kitchen table . . . the little, fleeting things that tie families and communities together.

None of the visits or invitations I'd hoped for ever really materialized, and unlike in the city, where there are always lonely people looking to make friends, everyone here was part of a group of friends they'd had since birth, and they didn't need anyone new. I wasn't a part of this ecosystem anymore, and I'd changed so much that I didn't readily fit back in, either. I no longer had the easy knowledge of who everyone was, and I wasn't a part of anyone's inner circle. But this was my home, and if I didn't belong here, I didn't belong anywhere. A disorienting, creeping sense of ennui set in; nothing was really identifiably wrong, but nothing felt quite right, either.

It was a scorching day in July when I learned I was pregnant. Years of musing about motherhood coalesced into flesh in the blink of an eye.

We'd only been married a month. I hadn't thought it would happen so soon; I had this vague notion that everyone seemed to have to try for half a year or so, and I'd planned accordingly. I was by myself at the farmhouse when I saw the faint blue line. Good. I wanted to hold this secret alone for a little while. To try to figure out how I felt about it. Without thinking, I put on my running shoes and began a slow jog down the ruler-straight, dusty gravel road we lived on. The drone of grasshoppers and the hot smell of crops baking in the sun were so thick they seemed to be the medium I was running through, while the corn rose up on either side of the road like a parted sea. The green glow from all sides soothed me, and the rhythm of running and the firm strike of my feet on the ground gave my body a way to redirect nervous energy so my mind could turn this new thought over.

A baby. Minuscule, but growing. Settling into my body and beginning to reshape me around itself.

I'm not alone in here anymore.

I pushed myself to run harder, as though it might help me to somehow force the truth into my head, making it seem real. In my fantasy world where pregnancy was just something we were slowly working toward, that second blue line was meant to come some hurried morning before I headed to work at a lab somewhere. Instead, it came amid a morning of searching scientific job ads filtered by how far I could possibly drive from our rented home. I was meant to have gotten my shit together and not be unemployed when this happened. I suddenly felt an urgent, panicked need to find my next position, popping the peaceful domestic bubble I'd been living in since we moved. I thought of Carole and promised myself that this baby wasn't going to break my focus. I would keep looking for a position regardless of what was going on in my abdomen.

As I turned back into our driveway, finishing a hairpin three-mile circuit, sweaty and dusty in equal measure, I saw that Eric had returned from his morning errands. I lay down on the lawn in the cool shadow of a silver maple, panting, and waited for him to come over. He sat on the ground next to me, looking relaxed. How long did I want to keep this secret? I'd thought I wanted it for myself for a few days, but he looked so calm and content, and the day was lovely and perfect. Eric's happy nature had always provoked openness from me; it was one of the things I loved about him.

"I'm pregnant," I said simply, looking up at him.

Unsurprised, somehow, he just smiled brightly back at me, then pulled me up into a hug. For Eric—schooling finished, career neatly lined up before him—it was an uncomplicated joy. He had never been conflicted about becoming a parent, and had never needed to ask himself how it would affect his body or his career. How I envied that.

I told him I was heading inside to take a shower. He nodded and kissed my neck. "You taste like the ocean. Go now, become a lake," he said in his oddly poetic way. "I'll make us lunch and we'll celebrate."

Plants are experts at repurposing parts of their bodies. Flowers, for instance. The very petals that make them so stunning and irresistible to pollinators evolved from leaves.[6] An organ that once turned sunlight into sugar became a fragile thing that exists only to act as a semaphore for a short time and then wither away. But what an amazing purpose they serve, making the species that bear them the most successful group of plants ever to evolve. Looking back even further, plant roots adapted their form to give space and protection to mycorrhizal fungi, a symbiotic partner that, in exchange for a home and a bit of carbon, enormously expanded the reach of those roots, allowing those that hosted them to better survive drought and poor soils and facilitating their takeover of terrestrial Earth. The ability to adapt, to transform parts of yourself into whatever the situation calls for in the moment, is what distinguishes success from failure, whether you're a plant or a person. But the cost can be high.

The days rolled on. A few weeks after learning I was pregnant, the rest of my body abruptly caught on, and I found myself nauseated and exhausted nearly every waking moment. Daily, I looked down at my stomach to see if anything looked different, but nothing seemed to have changed there. I found it hard to believe in a being whose only discernable effect was to make me feel gross. It all seemed so unreal. I did little during those first queasy months, when Eric would disappear to work early each morning with our only vehicle and I was left to fill the day alone, unable to go further than I could make it on foot, which was only down the same dusty road amid the rows of corn and soybeans. Still, I made myself check job boards and follow leads nearly every day.

This wasn't where I expected to find myself half a year after finishing my PhD. The farmhouse was limbo made real. All of my research notes, microscope slides, textbooks—all the things that reminded me of who I'd been before I left Montreal—hibernated in boxes around the house, lost among the other things I should unpack if I could only find the energy. Without a focal point for my restless mind, a goal to pursue, the silence felt suffocating.

My nausea ebbed with the heat of the summer, and as the last of the soybeans were harvested and the leaves crisped on the ground, my stomach finally thickened and started to round out. Each day, gravity pulled me down a little harder, and every night, I lay awake with the insomnia that would stay with me through the rest of the pregnancy. In bed, I listened to the soft, distant noise of the late-night combines combing the fields, just as I had when I was a kid. I felt comforted by the familiarity of this very specific sound and time of year. I tried to reconcile myself to the fact that soon, the bump I'd waited for would render me unhireable. If I wanted to start a postdoc, it was either soon or not for a long time. It seemed likely to be the latter.

I was more than a little bit surprised when an email arrived inviting me to interview for a postdoctoral position at a government research facility nearby. This interview might be my only chance.

The project was in trying to better understand the genes that control plant root structure. I'd be working in the lab to isolate and sequence the products of those genes, called transcripts—little bits of RNA that act as the messenger between the DNA and the protein-assembly machinery that builds what the DNA tells it to. I'd then take the massive amounts of data produced by the sequencing and write programs to analyze how the sum of all those transcripts, called the transcriptome, changed when certain genes were knocked out.

It was November by the time I walked into the research facility and met my would-be supervisor for the first time.

"Dr. Zimmerman! Pleased to meet you," he boomed with a smile and a hearty handshake. He was a friendly and jovial middle-aged man. I was at the very outside edge of being able to hide my pregnancy but pulled it off with a drapey shirt that flared around the waist. I wanted to get through the interview without my "condition" being front and centre. We sat in his office and talked about the project. He was unconcerned with my lack of a background in transcriptomics, saying I could pick it up as I went. Still, I walked out feeling certain I wouldn't hear from him again.

The afternoon he called me and offered me the position, I was washing my first sets of newborn clothes and feeling a bit stunned at how tiny they were. My abdomen had rounded out further in the short time since the interview and was now undisguisably pregnant. I wasn't obliged to tell him, but I didn't want to get off on the wrong foot by withholding this information. Better he know now. So I told him that I was expecting my first child and would give birth only a few months after the appointment was set to begin. To my surprise, he didn't seem bothered. My fellowship only allowed for a four-month maternity leave anyway, he told me, and there was plenty of work I could do while I was at home with the baby.

To American readers, a four-month maternity leave will perhaps seem generous; plenty of time to recover from childbirth and get back on one's feet again. For Canadians, among others, this will seem appallingly short, allowing precious little time for bonding or breastfeeding.

And indeed, this was the reaction of most of my friends and family, who were lucky enough to take for granted that mothers in Canada get at least a year to be at home with their babies. Upon learning of it, many assumed I'd just misunderstood, and that a year of leave is guaranteed for all Canadians; they were scandalized to learn otherwise. What they didn't realize, and what I was just coming to understand, is that jobs can be creatively categorized in a way that does not guarantee even a researcher working for the government itself the benefit that nearly all other Canadians receive. Whether four months seems to you to be a generous allotment or an absolute travesty, the takeaway lesson is that those in power will quietly give mothers less if they are able to, and that being a highly educated professional doesn't change that.

The same week I accepted the job, I hit the halfway point of my pregnancy and had my anatomy ultrasound. A phone call from my midwife the next morning confirmed that the baby was healthy and that it was a girl. Alone in the farmhouse, as usual, I hung up the phone and slid down the doorjamb, sitting on the floor in the sunny kitchen. A daughter. The sense of reality that had escaped me for months finally arrived with a bang. That I, who had grown up a little girl without a mother, was going to be mother to a little girl. It felt like a chance to fix something that had been broken for a long time.

In the space of a few days, it seemed like my life had turned around. I could think happily about the daughter I would have and know that she would have a scientist for a mother. With a goal I could now firmly latch on to, I immediately hit the books, bent on using the two months I had left before my postdoc started to learn everything I could about it and get a running start from day one. If I just worked hard enough, I could still make it all come together.

CHAPTER 8

THE LADY VANISHES

I certainly think that women though generally superior to men [in] moral qualities are inferior intellectually; & there seems to me to be a great difficulty from the laws of inheritance, (*if I understand these laws rightly*) in their becoming the intellectual equals of man.

—CHARLES DARWIN, **Remarks on the intelligence of women in a letter to C.A. Kennard, January 9, 1882**[1]

In the special case of science and engineering, there are issues of intrinsic aptitude, and particularly of the variability of aptitude, and that those considerations are reinforced by what are in fact lesser factors involving socialization and continuing discrimination.

—LAWRENCE SUMMERS, **Harvard president, Remarks on women in science at the National Bureau of Economic Research, January 14, 2005**[2]

> Let me tell you about my trouble with girls . . . three things happen when they are in the lab . . . You fall in love with them, they fall in love with you and when you criticise them, they cry.
>
> —**TIM HUNT, Biochemist and Nobel laureate, Remarks on hiring women in research, World Conference of Science Journalists in Seoul, South Korea, June 8, 2015**[3]

The lab was brightly lit and full of its usual lines of conversation, crossing back and forth among the half dozen or so lab techs, postdocs, and graduate students working at different tasks around the room. It was a typical-looking molecular biology lab, with black-topped workbenches running along the walls down the length of the room on both sides and a central bench down the middle. Gas and vacuum nozzles jutted up here and there. Above the benches were shelves covered in orange-lidded chemical reagent bottles of all sizes, as well as pipettes, beakers, and boxes of latex and nitrile gloves. At the back of the lab was the door to the supervisor's office; not since my time working for Larry had I had an advisor whose office was adjoined to the lab, and I was hopeful that it signalled a similar level of warmth and cohesion within the group. As I was the most recent addition to the team, my office space was in a lonely room on the far side of the facility, a several minute walk away if you didn't want to take a shortcut outside through the Canadian winter. It was inconvenient, but I knew I focussed better without the background noise of multiple ongoing conversations, so I didn't mind. I was in the lab daily doing experimental work anyway, so I could still be a part of the group that way.

I'd been officially working as a postdoc for about a month, and had started to get into the routine of my new position. My first weeks there had been overwhelming as I pushed myself to read all the papers required to get up to speed on the project, to learn where dozens of tools, pieces of equipment, and chemicals were in a new lab and a new facility, and to absorb both the concepts and the techniques required to work in a new field. I had essentially moved to a new branch of science and was trying to go from zero to competent as fast as I could. What would have been a steep learning curve for anyone felt just a bit steeper each day as my energy waned. I was seven months pregnant, and my white lab coat flared out like a tent as I waddled around fetching reagents and glassware for my experiments. Being on my feet for much of the day drained me physically and the massive information uptake drained me mentally, but I felt things were going reasonably well. That said, the information uptake was enormous, and not everything stuck the first time. Some days, I had to ask for help, and today was one of those days.

"Brenda, can you remind me where the cryogenic gloves are?" I asked the senior lab tech, a knowledgeable but sarcastic woman in her fifties whose mercurial mood had to be cautiously assessed upon entering the lab each day.

"I told you where they were already. Geez, that baby's really fogging up your brain, isn't it?" she said with a cool smirk.

"Just wait until the baby's actually here. We'll be telling her how to do her experiments. Or doing them for her," said Tracey, another lab tech of about the same age who projected a mixture of grumpiness and indifference, but who was at least consistent and predictable about it. She was grinning smugly. They both looked for my reaction.

I knew well enough to not respond and to just let them have their laugh—reacting would just bring accusations of having no sense of humour. I gave a weak smile and went to get the gloves for working at very low temperatures, but I could feel my cheeks flush hot with embarrassment. Having the senior lab members taunt me about my slips of memory made me feel like an incompetent moron. I shuffled my way back to my bench space and perched my expansive body on a stool that made my back immediately ache, determined to show how quickly and precisely I could carry out the protocol for extracting mRNA, because

I suspected they were paying attention. Little digs at my memory and competence weren't something I'd experienced in graduate school, but they were par for the course during my postdoc, where my body now seemed to invite both comment and mild derision.

When I got the word that I'd been hired at the research facility, Eric and I had reluctantly given up the farmhouse we'd been renting and moved into the city. He was working part-time at two different optometry practices that were over an hour's drive apart and we only had our old station wagon to get around in, so continuing to live in the country wasn't going to work. After missing rural life for so long, giving it up again after less than a year was hard. We were once again in a small apartment, this time in the bottom half of an old house in the sketchy part of town. Eric would drop me off early in the morning and pick me up again well after the winter sun had set and we'd finished our long days. On Fridays when he'd pick me up, we'd drive to a nearby, nearly always deserted Dairy Queen and eat a Blizzard to celebrate the end of another week while we sat in the glare and flicker of the fluorescent lights and stared out at the snowy darkness. We imagined what our daughter would be like, made plans for the future, and cast exhausted smiles at one another. This wasn't quite the romantic first year of marriage I'd pictured, but we managed to find the sweetness in small moments. We promised ourselves that one day we'd escape the city again and find ourselves another farmhouse.

My postdoctoral research involved working to better understand the functions of certain genes related to root structure. I'd first grow a large crop of study plants. Some of these had been altered at the genetic level so that the genes of interest didn't function, and I could learn about them by seeing what happened to the plant in their absence. Some of the plants were unaltered—what we call wild-type plants. These latter served as a sort of control group for comparison. Once they'd reached a certain age, I extracted their messenger RNA (mRNA) and sent it for sequencing. The big difference between the DNA I'd worked with during my PhD and mRNA is that DNA is *vastly* more stable. DNA, so long as it's kept in the right conditions and not exposed to certain chemicals that destroy it, can last for a *very* long time. The oldest DNA successfully sequenced is more than a million years old.[4] Conversely, mRNA simply conveys the information in the DNA to the machinery of the cell and then breaks down. It is by its very nature ephemeral. So to work with it in the lab you have to be quick, and you have to work at a very low temperature in liquid nitrogen, which slows the decay. These aspects made it fascinating, though altogether more stressful, to work with.

Once the mRNA had been sequenced, I took the thousands upon thousands of short sequences I got back and wrote programs that analysed the overall changes caused by knocking out the study genes, either alone or in tandem. Sequencing the total mRNA gives a sort of instantaneous snapshot of the genetic activity of the cells so you can see any downstream effects caused by a gene not functioning. This is less obvious than you might imagine because many genes don't have only a single, straightforward function. Knocking out some genes can have myriad, seemingly unrelated effects throughout the plant.

I didn't mind the lab work, but what really fascinated me was the analysis. Biology can be slippery. There are exceptions to practically every rule and very little is as cut-and-dry as you'd like it to be. Like gene functions or the dividing lines between species, as we've seen. I sometimes missed the black-and-white certainty I felt during my math-filled days as a physics major. There's a comfort is being able to snip the world into neat little pieces. Writing bioinformatic programs gave me a taste of that old clarity. All the biological complexity of the genes I was working with got boiled down to sets of sequences that either were or were not present in the cell under different circumstances.

My life felt like so much shifting sand under my feet: changing homes, learning a new field, awaiting a child . . . Everything was uncertain. But here I could sit down at my desk and deal in absolutes. What's more, getting to sit and think about a problem, focussing my whole mind on the task at hand until I could see a way to tell the computer how to extract an answer from that wall of data, was profoundly satisfying and gave me a real sense of accomplishment. I missed working in natural history, but this part of the project, at least, I could enjoy.

Other aspects of my position, however, weren't so simple, thanks to my pregnancy. The lab maintained a plot of experimental plants in the facility's greenhouse. Not ones I worked with directly, but all lab members were expected to pitch in and help care for them by watering, repotting, and doing other maintenance as needed. It's the nature of greenhouses that, sooner or later, they will develop pest problems that need to be controlled, and this one was no different. The issue was dealt with by regularly spraying pesticides. The plants were considered safe to be around a day or two after spraying, and workers wore lab coats, gloves, and maybe a light dust mask. As someone carrying a fetus around, I wasn't so confident in the safety of plants covered in pesticide residue, so I asked to contribute in some other way than working in the greenhouse. I was told not to worry about it, that it was safe once the waiting period had elapsed. The dismissive attitude annoyed me. When I kept pushing, I was told that if I was concerned, I could simply wear extra protective equipment into the greenhouse. Unfortunately, that amounted to having to work in not just the usual gear but a heavy-duty face mask and shield that, combined with the heat and humidity of the greenhouse, left me dizzy, panting, and exhausted by the time I finished my work. I was frustrated at both the disinterest in my concerns and the refusal to allow me to contribute in a way that wasn't so arduous.

At other times, the work was unavoidably hard on my pregnant body. All of my experimental plants had to be harvested and their RNA extracted at precisely the same age, which meant that to begin, a huge number of seeds had to be started growing on the same day. This involved surface-sterilizing the small seeds, painstakingly rubbing each one on sandpaper to abrade the seed coat for faster germination, and laying them out ten or so at a time on wet filter paper inside Petri dishes, which were then sealed. Everything had to be autoclaved and kept sterile for the whole procedure, which meant a *lot* of careful preparation followed by lots of hours working in a laminar flow hood. A flow hood is like an enclosed desk with just the front open so you can reach in, while sterile air is constantly being blown at you so that none of the nonsterile air of the surroundings can get in and contaminate your samples. The air around us is surprisingly full of bacteria and fungal spores, so for procedures demanding a high level of sterility, even normal air will ruin the procedure. Seed plating day was the sort of thing you saw coming and planned for weeks ahead of time because it was going to be a very long day at the lab and everything had to be ready.

The day I had to start the majority of the plants I'd be working with in my research, amounting to hundreds of seeds, was the day I learned that you cannot sit naturally at a laminar flow hood while seven months pregnant. You have to reach into the hood, and your belly is very much in the way. After contorting myself in every way possible to try to make it work, I found all I could do was press my belly up against the hard edge of the desk and sort of bend over it. Within ten minutes, an ache followed by pain bloomed up my lower back and hugged around my ribs. I tried facing sideways and just twisting my upper body, but that was worse. There was nothing to do but bear it, try to work quickly, and take frequent breaks, which just made the job take longer. I started early, but by the time I finished, it was eight o'clock in the evening and dark outside. Everyone else had gone home. My back was so painful that it spasmed when I slid off the chair, and I had to limp my way, hunched over, to the growth chambers to set out my Petri dishes of seeds. I couldn't straighten.

That evening, I lay curled on the couch in pain, holding a warm rice bag to my ribs and another against my lower back while Eric massaged my knotted muscles to try to loosen things up. The baby, blissfully unaware of the position we'd spent the day in, didn't appreciate all the extra warmth and squirmed around restlessly. Her movements, now developed enough that I could tell where her head, bottom, and legs were, made me smile, made the pain seem a little easier to bear. I whispered an apology to her for the heat and straggled off to bed.

But the work had needed to be done that day, and I'd finished it. Timing was paramount to the experiments. A few weeks later, I'd have a similar day planned for the harvest of those same seeds—another long and involved procedure—before being hit with a stomach flu so bad that I had to go to the emergency room and receive intravenous fluids to prevent dehydration that could cause preterm labour. When I sent my supervisor an email letting him know that I was being kept at the hospital for observation and wouldn't be able to harvest the plants on the planned day, his sole response was to ask me to confirm the harvest date so he could make sure it got done by someone else. While I was relieved that weeks of work and all that seed plating hadn't gone to waste, the curtness of the message stung.

By late March, with only a couple of weeks left before my leave was due to start, the days had reached the outer limits of my ability to push through them. I woke exhausted from poor sleep and dragged myself through the day in a near-fugue state, everything done as if in slow motion. I struggled to stay awake at my desk and wheeled metal canisters of liquid nitrogen down the hall like a haggard hospital patient meandering down a hallway with an IV stand—one step at a time, looking like I was about to drop. By the time my leave date finally rolled around, I was running on fumes and thankful to just get off my feet and rest.

While leave came as an incredible relief, I didn't intend to just let my work drop. I knew my life was about to change, and I wanted to be ready. I bought books about managing life as a working mother, and one called

How to Succeed as a Scientist: From Postdoc to Professor. My supervisor had sent me a bunch of research papers he'd recently found and thought might be useful so that I could read them while I was off. He also wanted me to get some writing done for the eventual publication of my research. I had to be off my feet, but I didn't have to stop working; I could use this time to come back better. That was what I was hearing from him, but I'd also bought into the idea that I should be working no matter what.

I had no idea of the collision that was about to happen between my personal and professional lives, and the damage it would do to both.

Postdoctoral research often falls at a particularly complex time in a woman's life. It's a key juncture in what's been referred to as the "leaky pipeline"[5] in which highly educated women increasingly opt out of academic careers as they progress through the ranks, resulting in few reaching senior roles, despite strong numbers at earlier career stages. Several factors collide to create the perfect storm.

Postdoc researchers are working in demanding, competitive positions that tend to be poorly paid, sometimes without benefits, and that carry no guarantee of attaining the next rung on the ladder. The whole enterprise of postdoctoral work is massively lacking in stability, both because it is short-term contract work and because finding the next position after the current one ends can require moving across or even out of the country. Each successive position will entail starting from square one in a new environment with new colleagues, and often at least somewhat new subject matter. It is a nomadic existence. Even for those who successfully complete a string of postdoctoral research positions, there is no guarantee of a tenure-track job at the end of it. In fact, beyond a certain point, it's a commonly held belief that a greater number of years spent as a postdoc can make you *less* likely to be hired because it starts to look as though you don't measure up. So there's enormous pressure to publish work quickly and land a faculty position, lest you face diminishing returns.

Now add in the fact that women are navigating this gauntlet at the same time of life they are typically starting their families. Women are still the primary caregivers for children in most heterosexual marriages and will be expected to fulfill that role even while holding down a challenging research position. For many, knowing what they're up against leads them to a decision to opt out of pursuing the tenure track before they even start that family.

According to a large 2011 meta-study of researchers and trainees in the academic science pipeline, the mere *intention* to have children in the future leads to an 8 percent increase in dropout rates for female postdocs versus their childfree female colleagues, while *actually having* a child before or during the postdoc leads to increases of 12 and 21 percent, respectively. Over 40 percent of surveyed postdocs who had a child during their postdoc shifted their career goals away from academic research. For women, issues surrounding children were the top-cited reasons for abandoning those career plans. For male postdocs, there is no such effect from having children.[6]

The hiring numbers bear out those women's fears: female researchers with young children were found to be 33 percent less likely to land a tenure-track position than those without children, and 35 percent less likely than men *with* young children, who are at an advantage merely by being men.[7] For those who do continue on in research after having become mothers, the productivity penalty is significant. A 2018 survey of more than three thousand tenure-track faculty members from across the US and Canada found that in the five years leading up to the birth of a first child, the male and female parents' research productivity was identical, but following a birth, the mothers' output immediately dropped by up to 48 percent. There was no corresponding drop for fathers, allowing them to pull ahead and create a gap that, by the time the children in question turned ten, would take the mothers a full five years of extra work to close.[8]

Renate Ysseldyk, associate professor in the department of health sciences at Carleton University, has both studied the struggles of female postdocs and experienced them for herself when she had her first child as a postdoc. "It's hard, because suddenly you have these two potentially very conflicting roles," she says. "Academia in general is not necessarily very kind to caregivers. The expectation is that you're always working. There's this unwritten rule that you just have to be always on and producing. And so trying to reconcile, well, am I a mother or am I a scientist? Can I be both at the same time? It's something I still struggle with, and my kids are not babies anymore. Parenting is a full-time job, too."

Though Ysseldyk ultimately chose to remain in academia, she admits that wasn't always a certainty. "I'm very grateful that I actually managed to get an academic position, but for a long time, I wasn't sure that it was going to happen. I simultaneously felt like a negligent mother and a career failure. I remember standing in my kitchen crying, talking to my husband and saying those words and thinking, 'I don't think this is ever going to happen for me.' I was very close to walking away."

In a study published in 2019, Ysseldyk and her coauthors interviewed postdoctoral women in Europe and North America working across a range of fields about their conflicting identities, feelings of control or lack thereof over their lives, mental health, and the gender-based challenges they face. Overall, the women reported low levels of satisfaction with their lives and only moderate levels of perceived control over their careers and their futures. They felt that it was looked down upon for women to have kids while doing their postdocs, and that maternity leave would cause them to fall behind their peers. Even after the leave had ended, the perception was that it's very difficult to compete with those who did not have children. As one woman in the study put it, "You are either expected to play the game in full or get out."[9] It's no surprise, then, that the women reported mental health issues that included depression and anxiety, often made all the more difficult by how tightly entwined their work and personal identity are. They also noted that the highly competitive environment in which their work took place hampered female researchers from supporting each other in any meaningful way.

Ysseldyk's study found that the difficulties women face in the STEM pipeline tend to be cumulative, and that often the postdoctoral stage represented a tipping point at which many women felt the sacrifice stopped being worth it. "There's a lot of uncertainty about whether or not that full-time academic position is ever going to materialize. And so of course, that's when many say, 'I've had enough of this. I need to go do something else,'" she explains. "It's not this one big, dramatic event—it's all these small things put together. And then maybe there's a moment that's like, 'Okay, this is the last straw.'"

Ysseldyk points to faculty supervisors as a key help or hindrance for postdoctoral women. "I think [supervisors] have a big responsibility to be informed about research like this and the inequalities that do exist, because there are some [who], whether knowingly or unknowingly, end up sabotaging their postdocs' careers by expecting too much from them. I think it's important that [supervisors] are supportive," she says. "A good starting point is for [them] to educate themselves on the challenges that postdocs face, especially as women at this life stage." In Ysseldyk's own case, she says, good mentors may have made the difference between staying in research or jumping ship. "Without those mentors, I'm not sure I would've lasted even those first few years in academia."

At a 2021 virtual conference organized by the nonprofit organization Mothers in Science, cofounder and chief executive Isabel Torres highlighted the fact that the obstacles facing mothers are to a large extent invisible, leading the women themselves, as well as those around them, to assume that all it takes to succeed is hard work and determination. Because of this perception, when a woman finally chooses to walk away due to the fallout of systemic failings, she said, "it looks like a personal decision."[10]

Having started my maternity leave just a couple of days before my actual due date, I then sat and watched with increasing dismay as nearly two weeks of that leave went by without anything to show for it. Warned as I had been about the dangers of premature birth, I'd been entirely unprepared for the possibility that the baby would be *late*. I sat in our apartment—which, despite its small size, had suddenly begun to seem underpopulated—and grew increasingly frustrated with my body and its refusal to adhere to my work schedule. I had what felt like so little time to be at home with this baby, and now I'd spent a significant portion of it empty-handed, struggling to keep my mind on my reading while it kept pulling away to scan my body for any change in status.

On a sunny Sunday afternoon in mid-April, it finally happened. A low, blooming cramp in my lower belly that came and went in what gradually coalesced into a pattern—growing, peaking, and ebbing away—that took a little under ten minutes to repeat. It wasn't particularly painful yet, and I was practically giddy with joy that things were finally happening. I called the midwife to give her a heads-up that she might soon have work to do, and I was told to get some sleep if possible, and to do plenty of walking to help keep things going. The plan was to have the baby at home; moving into the city had given us quick access to a major hospital, so I felt confident giving it a try, knowing I wasn't far from help if it was needed. I was too excited to sleep, so I walked around the neighbourhood and then around the apartment that evening, settling into a *Police Academy* marathon with Eric once I was too tired to keep circling the living room. I barely slept that night, being too tired to pace but too uncomfortable to drift off. The contractions had certainly started to hurt, but never seemed to get any longer or closer together, as I'd been told they would. When nothing had changed by the next morning, I was getting discouraged.

Little did I know just how badly my body seemed to want to hold on to this baby. More than twenty-four hours after the contractions had started, the midwife gently informed me that the labour wasn't progressing. She worried that if this went on too long, I'd tire and not have enough energy in reserve to sustain me through the birth. We made the decision to head into the hospital. I was disappointed but worn out, and just wanted things to get moving at this point. At ten days overdue, it felt like I'd waited forever. I would eventually learn that this is just the way with my body and my babies—I never let go of them without a bit of chemical and/or mechanical prodding.

I did my best, with Eric's help, to gather what I needed and get a little more decently dressed for my trip to the hospital, and we were on our way. The change of pace must have made a difference, because around the time we got to the hospital, the pain abruptly got much more intense. This wasn't *excited-because-something's-happening* pain, or even *grit-your-teeth-and-bear-it* pain. This was scary, *make-it-stop* pain. This was a ten out of ten. I tried to think about Lyn, and the Litany Against Fear that she'd copied into her journal. I tried to let the fear pass over and through me. But now, when it really mattered, I couldn't.

That low, blooming cramp now felt like what I imagine getting stabbed would be like. I crumpled around my hardened belly and struggled to breathe fully each time a peak hit me, further twisting the knife. A hot bath, followed by nitrous oxide, followed by a first failed attempt at an epidural—painful in itself, but nothing compared to the contractions—had not helped, and I felt like I was losing my mind. I could hear the agonized noises I was making as if from a distance as I squeezed my eyes shut against the pain. Nothing anyone said had any meaning anymore. I'd later find out that the baby was tangled in her cord, and I was contracting futilely against something that couldn't be moved. A more senior anesthesiologist was brought in for the second epidural attempt and, as if in apology for the added suffering, hit me with so many drugs I lost all feeling below breast level almost instantly. I'm certain that nothing short of dropping dead will ever bring me such a deep and profound relief from such acute pain as that epidural did. It was bliss. People can be pretty competitive about refusing epidurals, as though suffering is something you get a trophy for after the birth, but for me at least, getting one meant I got to experience childbirth as a rational human being able to form long-term memories, rather than a half-crazed animal in a full panic. I got to be present.

After myriad procedures meant to monitor and untangle the baby, including pumping in artificial amniotic fluid to take some pressure off the umbilical cord—procedures that I could now watch calmly with the scientific curiosity of a biologist attending her first human birth—it was finally time to push. The baby wasn't doing all that well after being contracted against for more than a day and a half, the doctor told me, so this needed to happen quickly, or else it was surgery time. After everything we'd been through, the last thing I wanted was to also get sliced open, so I gave it my all. She was born in such a rush that the resident had to scramble to catch her.

And there she was. This person I knew intimately, but had never set eyes on.

Clementine.

We'd chosen the name months ago, tentatively using it between us to feel it on our tongues, imagining what it would be like to have such a person in the world. And here she was. My Clementine. The nurse laid her out on my chest in just her diaper and a pink toque for that first skin-to-skin time. She was scrawny, on the small side of average, and seemed totally powerless in the face of gravity. She just laid her head on my chest and stared out, wide-eyed, at a new world. I'm not sure I'd ever even held a newborn before that moment, and I was dazed, looking at her tiny body as though from a distance and noticing vaguely how little she weighed in my arms. It didn't seem real. Then, with massive effort, she turned her head, opened her tiny mouth, and started gently searching with it across my chest, looking for a place to feed. I was struck dumb with her utter dependence on me. Her helplessness. The conviction that she thought I knew what I was doing, and that I would do anything not to fail her. Things changed for me in that moment, so certain was I, all of a sudden, that I *would* do absolutely anything for this messy and confused little ball of life. This thing needed me, and I wasn't going to let her down. Behind that, the ghost of a thought . . . *Is this the only living thing I'll ever give a name to?*

The early days of motherhood passed, as they always do, like a fever dream—half-remembered moments of sweetness shot through with sleepless nights and a sense of being unmoored from time and reality. The world barely existed outside our home. Pushing through weeks of sore, bleeding nipples and awkward latches, Clementine and I settled into a rhythm and understanding of each of our roles in feeding that led my scrawny, plucked chicken of a baby to blossom into a plump-cheeked, more human–looking creature. The rate of change was astonishing, yet I was never able to perceive it except in looking back at photos of a face I hadn't yet realized was gone. I'd named her after a character in a movie who'd wanted to forget someone she'd loved, and it felt painfully ironic that I was now struggling to remember her face.

Sitting on a park bench one day and peering into the stroller at her sleeping form, I pulled out my sketchbook and pencil. I now understood Lyn's words about infinitesimally small changes adding up to a profoundly different being. Maybe if I drew her, taking in each curve and shadow in turn, they would imprint themselves on my memory the way my plants did, and I'd be able to remember this version of her. A perfect moment in time, when she was about two months old. I already couldn't clearly recall any of her previous versions, and I couldn't bear to lose yet another one. It was all going too fast. I put pencil to paper and tried not to think about all the future versions that would slip away in turn.

As summer came, Clementine was just gaining some decent head control and had emerged from the early tempest of colic, revealing herself to be a happy, playful baby with an easy laugh and a go-along-to-get-along attitude toward life. I was completely charmed by her, this happy little person for whom the world was new. Her wide-eyed curiosity made it new for me, too—made me look again at things I'd forgotten were wonderful and begin to notice once more. Was this why people loved parenthood so much? Maybe it wasn't just slogging through the early years to get to the fun of an older child. Perhaps this phase, too, was its own reward. In our brief pocket of time together, the days simultaneously stretched out forever and fell away too quickly. I spent as much time as I could just holding and watching her, trying to cement these moments in my mind. I imagined that I must have had days like this with my own mother. That she must have once looked at me and loved me in a way that made her heart ache. What must it be like to grow up basking in that warmth? I hadn't understood that it would be like this.

As blissed out on motherhood as I was, I was determined to go back to the lab and be, in Larry's words, "the best," still certain I could do it all. I'd been reading my book on how to succeed as a junior researcher and brushing up on the newest scientific literature surrounding my project—anything I could do while my daughter slept that might better my chances of success when I got back to work. Always, I did this with an awareness of the fear and anticipated pain of leaving her every day at such a young age, the anxiety of watching the all-too-short days with her slip away. I just wanted more time. Four months at home with her had seemed totally reasonable on paper, but I now realized I'd agreed to this without knowing what I was getting into. I felt like I'd made a deal with the devil—greater research productivity and career success in exchange for my firstborn child, delivered to any day care at fourteen weeks of age.

I was trying to figure out how best to arrange my working day around a single car that also had to get my husband to work and back and my daughter to day care and back within their rigid schedule. With no feasible public transit options, I figured out that working a base day of seven to three instead of nine to five, with later hours when needed, would maximize my working day while meeting all of the other requirements. A two-hour shift in working hours for a postdoc who works largely independently on a project is generally not even noticed, much less problematic. One of the perks of being a poorly paid academic is usually a great deal of flexibility in one's working hours, something I'd learned to enjoy during my PhD. Nevertheless, I thought I should get in touch with my supervisor to let him know what my plans were and that I was thinking ahead to being a good worker when I got back.

His response was a reminder that I needed to lead the project, not just show up to work. It was a mild rebuke, really, and he did ultimately agree to the time shift. But I had just spent a significant chunk of my all-too-short time home with my daughter studying how to be the best possible project leader, and this felt like a blow. The mere mention of a modified schedule designed to accommodate parenthood seemed to him to necessitate a lecture on my responsibility to my work. I hadn't even returned yet, and already my dedication was being questioned. I felt my cheeks flush hot with shame as I read his words and felt my enthusiasm dampen. As August arrived, I steeled myself for going back to work and leaving my daughter, revelling in the long, empty moments of just being with her and doing nothing, knowing those chances would end soon.

I arrived at the lab my first day with nerves already frayed. Clementine had been inconsolable at being brought to a new place and left with strangers, and I'd walked out of the day care feeling like all the life had been sucked out of me. I took a deep breath and willed my thoughts to my work. I wanted to hit the ground running my first week, feeling that if I could get some early wins under my belt, it would be easier to adjust. Getting the project moving again required a meeting with my supervisor to touch base on where we were and what direction it would be best to move in; I couldn't really do anything until I talked to him, so after I set my things down in my office, I went straight to the lab to see him, hoping the bags under my eyes weren't showing too much.

Walking into the lab, I felt like a different person than the one who'd waved goodbye four months ago. I'd changed, fundamentally. But, of course, that wasn't obvious. All that was obvious was that there was a lot less of me than at any other time since I'd been there. Alice, a kind, soft-spoken Chinese postdoc who'd started at the same time as me, saw me and smiled.

"Hey, you're back! I feel like you just left. How are you? How is your baby doing?"

I smiled back and gave the standard answer, because it was impossible to explain to this fresh-faced and childless twentysomething that I was viscerally exhausted and felt like a bundle of frayed nerves stuffed into a body I didn't recognize and clothes that didn't fit. I asked her how she'd been and made small talk about the lab for a few minutes before looking toward the supervisor's office and asking where he was. Her brow furrowed slightly.

"He didn't tell you? He's away this week. He's not back until next Tuesday." A pause. More brow furrowing. "Maybe he forgot you were coming back."

" . . . Oh."

My heart dropped into my stomach. I didn't know what else to say, so I turned on my heel and walked out of the lab, making a beeline back to my office. Alice was almost certainly right. Despite our having talked just a couple of weeks prior, he'd forgotten I was coming back. I couldn't start back into my experiments until he and I made plans together. All I could do was sit and read scientific papers to stay on top of the research. And now I had put Clementine into day care and lost a week with my daughter so that I could sit here accomplishing nothing in a silent, empty office.

I sat at my desk, put my head down, and cried.

When he did turn up a week later and saw me approaching his office door, his first words were "Oh, I forgot you were coming back!"

I felt myself go cold. I wanted to snap at him. *Of course I'm back! I*

HAD to be back! But thanks to you, I've been sitting at my desk for a week doing things I could have done at home! But I was too well trained to let the anger show. I pasted on a wan smile and asked if we could discuss what needed to be discussed in order to get the project going. I just wanted to get through the meeting and get to work.

Soon things were up and rolling again, and I hoped they'd settle into a viable routine. And they did, for a while. The mornings were early and the days were long, but Clementine adapted to day care, and my work advanced at a decent pace with the generous daily application of caffeine from the giant thermos I became known for at the lab. But as summer turned to autumn and then winter, another pattern emerged. Every few weeks, Clementine would pick up some new respiratory virus or stomach bug at day care and have to be sent home, a stressor familiar to all working parents, and one that became much worse for a time in the wake of COVID-19, to the detriment of mothers in particular. Eric always had patients he had to see, and despite my long move home, we had no help from family, so it would automatically fall to me to be home with her, which of course meant letting my supervisor know that I would be missing a couple of days at work. My programming could be done on my government-issued laptop at home, but he was skeptical of the whole idea of working from home. He seemed to feel that people couldn't be trusted to work well without on-site supervision. This meant that each time I had to tell him, in person or via email, that I'd be home with my sick child, I was treated to repeated reminders that I needed to be dedicated to my work, and that I was making a bad impression as a project leader by being absent so much. Then he'd say that he'd continue to allow this because he'd agreed to be supportive of my role as a mother, seemingly patting himself on the back for being such a nice guy. Before long, it started to feel like he thought he was doing me a favour by employing me. He talked sometimes about raising his now-adult children while building his career, how it had been challenging but he'd managed it. Yet in the next breath, he'd mention that his wife had stayed home with their kids when they were young. He never seemed aware of the irony.

I started to feel anxious and panicky each time Clementine woke up with a fever or I saw the day care's number calling on my phone. I constantly wondered which would be the phone call that got me fired. As much as possible, I tried never to mention my home or family life. I desperately wanted people to see me as a professional and not just a harried mother in over her head. I tried to work harder to make up for any lost time and to show that I was dedicated to the project, but it felt like the delicate balance I'd established was starting to disintegrate. I never felt like I was doing enough at work, but I was so exhausted, I didn't feel like I was doing a good job of being there for my daughter when I got home, either. I was trying my hardest and floundering on all fronts.

Keeping my family responsibilities hidden took on a new and more literal meaning when I needed to pump milk for my daughter at work. Asking the head of human resources where I could pump produced a blank look, as though this had never come up before, though I found that hard to believe, given how short postdoc maternity leaves were. Were the other postdocs smart enough to not have babies during their time there? Had they just chosen other feeding solutions? I never found out.

"What do you need?" she asked, her voice sounding guarded, as though I was about to ask for something outrageous. A clean, private space with a table and chair, I told her. Not much.

"You can have one of the shower stalls in the ladies' change room. No one uses them anyway," she said, as though that would allow her to carry on with her day. "You can ask maintenance for a chair to put in there."

Asked about a small table to put my supplies on, she waved her hand dismissively. "Go buy one. Find something cheap and we'll reimburse you."

Which is how I found myself in a dark, cramped, mildew-smelling shower stall with a tiny folding plastic table from Walmart to hold my bottles. I was always terrified the flimsy thing would tip over on the uneven tiles, spilling something I wouldn't be able to replace. As it was, I couldn't believe they thought this was a fit place for an infant's food to be handled, but I don't think that was ever really considered. More than once, a staff member would hear the pump and whip open the curtain to find out what the sound was, startling and embarrassing me, not to mention almost spilling the milk. I eventually rigged a coat hanger with a handwritten sign to hang over the curtain when I was there. Three times per day I tucked myself away there and hoped no one came into the change room. I felt humiliated and hidden. Something to be stashed away so no one had to think about me. Real scientists don't have to stop their research for babies—that seemed to be the message.

Though I'd initially attacked the project with enthusiasm, that drive was wilting. I was used to putting myself in positions where a lot was asked of me; that was okay. But to keep making my way up the ladder with no additional thought given to my happiness or comfort in the workplace was getting frustrating. I felt like after more than a decade in research, I'd earned some basic consideration. How had I once been worth six figures in doctoral scholarship money but now wasn't worth a clean pumping space? Every day, I sat in there and felt my resentment grow a little bit more. The snide remarks I'd receive when I needed to schedule lab work around pumping, or leave at a reasonable hour to pick up my child, added to that feeling.

I became the hermit postdoc. I slunk into my office, red-eyed and exhausted from lack of sleep, every morning at seven o'clock before anyone else had arrived. I did my lab work early in the day to avoid my lab mates as much as I could and ate lunch in my office while working on my bioinformatic program—the thing that made sense in my world—and trying to see a way out of what was starting to feel like an untenable situation. I felt I was driving myself through each day on pure determination, but it wasn't going to hold out forever. I thought about my sunny mornings in the herbarium in Montreal and felt a different kind of homesickness than I'd ever experienced, one not for a place but for a moment in time when I felt sure of what I was doing and why I was doing it.

I became fixated on the two aspects of my life I felt I could control: my pumping, the one thing I could do that made me feel like a good mother while I left my daughter for long days at work and returned too tired to play with her, and my data analysis, which provided the hard numbers and predictability I'd had so little of lately. As I got more tired and anxious, the two ran together in my mind. I started to count the ounces I'd pumped on each side at each pumping session each day, plotting them with the same program I was using on my genetic data. I felt like I was drowning, but as long as I could keep all the numbers where they should be, through whatever white-knuckled discipline it took, life would make sense and I'd be in control. I was more than disciplined, I was obsessed. I needed this to work or I'd have failed to care for my daughter on the most basic level, I told myself. It took a long time to look back and recognize my fixation on pumping milk for the acute anxiety symptom that it was. Research like Ysseldyk's suggests that female postdocs needing mental health care for issues stemming from their work is far from uncommon, but despite working for the government, my creatively categorized position came with no benefits, and we couldn't afford it on our own. Even if we could have, it's not as though I could have asked for more time off to go to appointments.

All this was exacerbated by the feeling of moving through my days in a body I didn't recognize as my own. Not since adolescence had I experienced such an abrupt and profound shift in my body. It wasn't just the extra softness around the middle; my core felt weaker and I was more top-heavy than before, leaving me prone to backaches. My gait had changed, making me just a little bit clumsier, and a tendency to vertigo that I'd developed during the pregnancy just never went away. It was subtle in a way that made me wonder if I was imagining things, but my whole body moved differently, and I felt a strange sort of disassociation from a body in which I'd once felt graceful and well controlled. It added to a sense that I was now living in a twilight zone where everything about my existence that I'd once taken for granted had become a distorted caricature of itself.

No one had told me to expect these things. I realized, and not for the last time, that in many ways, being motherless is harder in adulthood than it is in childhood, when everyone is still looking out for you.

Eric, meanwhile, saw what I was going through and tried his best to be an equal partner and an involved father, but he was struggling, too. Because he'd had to take jobs in two different cities to make enough to keep up with his student debt, his days were as long as mine. When he finally got home, there were patient charts to complete until late into the evening, while I sat nearby trying to catch up on my bottomless work reading. We were both so depleted that neither of us had anything left to give, and there was no one to ask for help. We only had one another, but we couldn't save each other.

I can't remember the exact moment when I knew I was done, but by the time the days started to lengthen noticeably and I hit my one-year mark at the lab, I knew I wouldn't be putting myself through another winter of this, and that I would never repeat the experience with the second child we now hoped to have.

BOTANY HAS A LONG HISTORY of not holding space for women, and even pushing them out when it becomes expedient to do so.[11] During the late eighteenth and early nineteenth centuries, women were very much at the heart of botany. The European Enlightenment produced a huge interest in scientific knowledge, and in natural history specifically, as a form of self-betterment. The study of life was still tightly tied to religion, as a way of knowing God, and was seen as a means of moral improvement. Botany was considered a wholesome and enriching activity for "the fairer sex," and middle- and upper-class girls were encouraged from childhood to take part in a variety of ways.

"There were those who collected, and there were those who drew and painted. There were those who were involved with herbal knowledge in different forms. And there were those who were writing; writing as teachers and putting together informal manuals for teaching their children about botany, and for slightly broader audiences as well," says Ann Shteir, professor emerita in women's studies at York University who studies women's history in botany. "Fieldwork was there for women. That is, going out, collecting, preserving, getting information about how to name things, maybe exchanging specimens with others." Those with a particular passion were able to read journals, use microscopes, and attend meetings of local botanical groups as well.

Women of the late eighteenth century had greater access to botany than to any other branch of science, and so botany came to be thought of as a feminine activity in the popular consciousness. An 1827 biography of Linnaeus himself described the man's teenage disinterest in what he considered to be a woman's pursuit.[12] Encouraged by his father, Linnaeus's first botany teacher was his older sister. The lines separating science and nonscience, profession and hobby, weren't as strictly drawn at the time, so the study of plant structure and classification sat comfortably alongside the more aesthetic pastimes of painting flowers or maintaining a fern collection in the parlour. Distinctions between amateur and professional, too, were neither so important nor so black and white as they are today.

Beginning in the 1830s, however, what were then called "philosophical" botanists[13]—those that focussed solely on the rigourous science of plants, a group composed primarily of middle-class white men—began an effort to professionalize botany in hopes of gaining more respect for the field and establishing funded professional positions within academic institutions. Because a subject so heavily undertaken by women couldn't be considered a consequential intellectual pursuit, a major step in this movement was to "defeminize" botany . . . to sever the historical association with women so that the men who did it would be taken seriously.

A major player in the push to defeminize botany was John Lindley, an English botanist and orchidologist who in his inaugural lecture as chair of botany at the new University College London in 1829 told his audience that "It has been very much the fashion of late years, in this country, to undervalue the importance of this science, and to consider it an amusement for ladies rather than an occupation for the serious thoughts of man." Thus began the stratification of the field, with women and hobbyists driven to a lower stratum dubbed "polite botany," and men and their "serious thoughts" occupying the top tier of what came to be known as "scientific botany."

Part of this change entailed moving botany from the field into the lab and shifting its focus from observation to experimentation. "There are hierarchies of knowledge that sharpened in the nineteenth century," says Shteir. "And a lot of it had to do with professionalizing fields. So what we now call the amateur approach got pushed to the side as being less important, less contributory than those that were arrived at through techniques of classification, data analysis, and experimentation that took place within institutions. The more informal approaches gave way to the more formal approaches to knowledge. And those were associated with experimental techniques rather than observational techniques." This immediately excluded most women, whose lives were still largely carried out in and around their homes. And of course, the hierarchy of experimentation over observation continues to this day.

Another bit of sleight of hand that helped the process along was to, rather than simply push women out altogether, define a separate, related sphere of activities and encircle female botany enthusiasts within it, giving them the illusion of socially sanctioned participation and acceptance.[14] They were welcome to learn about botany, to an extent, and were encouraged to teach their children about it in order to improve their minds. For those women wishing to take part in polite botany, Lindley produced a book called *Botany for Ladies*, which he described as an experiment in conveying scientific knowledge "in a simple and amusing form."[15]

"It was botany for women. It wasn't botany for everyone. And it was not botany for boys," says Shteir. "He wasn't against women learning . . . it was what they were learning it *for*, and what they should use that knowledge toward. He was not interested in training women to be professional botanists. It was for women to do the kind of work within their natural orbit as he saw it."

Those few women who did conduct scientific research through their own individual efforts often found their paths blocked by the need to present their findings before a scientific society in order to achieve recognition for them. Many scientific societies of the day simply did not allow women, while those that did often restricted them to acting at best as audiences and at worst as set dressing, as in the case of the British Association for the Advancement of Science (BAAS), which issued ladies' tickets with the intention that their presence would raise the tone and mannerliness of the club's social events. One president elect is quoted as saying that women's presence even as an audience at the reading of the scientific papers would turn the affair into a "dilettante meeting" rather than a serious scientific discussion[16] but added, "their presence at private parties is quite another thing—and at these I think the more ladies there are, the better."[17]

One of the most vocal, intellectually formidable, and persistent advocates for Victorian women's right to fully participate in science was the English botanist and suffragist Lydia Becker. Born in 1827, Becker was the eldest of fifteen children and, after the death of her mother when Becker was in her late twenties, was forced into the role of mother figure to that large family.[18] Becker chafed at the "intellectual vacuity" of the life society expected women to live, writing achingly of the slow death of curiosity and ambition that awaited many women when they, "after vain struggles against their destiny, sink at last into a weary kind of resigned apathy, and men say they are content. But no one can measure the pain that has been endured ere the yearnings for a wider and freer existence subside into deadened calm."[19]

With a lifelong interest in plants, Becker turned her scientific focus toward extensive and detailed observations of a strange phenomenon in which a fungus forced the female flowers of red campion, *Lychnis diurna*, to develop stamens and become hermaphrodites.[20] Her investigations led to an ongoing correspondence with Charles Darwin, who helped her to frame her observations in light of his evolutionary theories, though they disagreed on what was actually occurring in the flower. Becker's interpretation was ultimately the correct one, as she theorized that the fungus co-opted the flower's anthers for its own reproduction, replacing the pollen that would normally have developed there with its own spores and spreading them via the plant's own pollinators, a highly novel finding at the time.

Becker's research led her to consider that the seemingly fixed categories of male and female might not be as immutable as they first seemed.[21] At this time, women's intelligence was widely debated, mostly by the men for whom their inferior position provided a direct benefit. These highly regarded gentlemen of science and medicine used women's supposed lower intelligence as a justification for not allowing women a proper scientific education; their low turnout in scientific pursuits were then used, circularly, as evidence of their disinterest in the same. Becker, having managed to become one of the very few women to present research in the Botany and Zoology section of the BAAS with her fungus work, gave another talk in which she used her insights on the fluidity of sexual expression to suggest that human minds might exist on a spectrum ranging from highly male to highly female, and that a more masculine mind might exist within a woman, and vice versa. She accepted that human minds ran the gamut of intelligence and ability but refused to concede that these were tied to physical sex. And furthermore, she claimed, were girls and women to be provided with the same education and upbringing as their male counterparts, their achievements would be the same.

If Becker's first talk produced bemused skepticism, the second caused a complete uproar. The advent of a woman coming forward and using scientific arguments to suggest that the fairer sex might indeed possess an equally sharp intellect was outrageous. Becker was roasted in the press on both sides of the Atlantic; commenters simultaneously implied she was stupid and out of her depth while paradoxically also calling her unnatural and repulsive for possessing such a masculine mind. Some sought to humiliate her and imply that she herself was a hermaphrodite. Though publicly unflappable, realizing as she did that a woman must be without emotion to be taken seriously among men, she confessed in a private letter to having a "horror of newspapers" after the affair. Still, there were those who agreed with her ideas, and the uproar carried her sentiments further than they might otherwise have gone; the talk didn't fall entirely on deaf ears.

Darwin, however, initially quiet on matters of female intellectual ability, began increasingly to apply his theory to humans following the publication of *The Origin of Species*, leading to harmful speculation on his part about the intelligence of both women and people of colour bolstered by a new interest in biological determinism and in spite of his earlier support of girls' education.[22] At around this time, and perhaps related to his statements, Becker's correspondence with him ended, and she shifted away from using evolutionary arguments to support her advocacy of women. One can only imagine how disappointed she must have been in her scientific hero.

While Becker's work eventually moved away from botanical research and toward fighting for girls to receive a rigourous science education, she nevertheless correctly interpreted the fungal manipulation of plant sexual organs and was the first person to document the phenomenon. In 1848 she published a book, *Botany for Novices*, which distinguished itself by being aimed not specifically at male or female readers but at *all* novices, and which provided accurate information using the modern impersonal voice that was a rarity for female writers at the time. Notably, it was bylined not Lydia Ernestine Becker but simply L.E.B.

I TOLD MY SUPERVISOR I would be resigning about a month before my daughter's first birthday, two months before my contract would come up for renewal. I'd agonized over it alone in my office for months, writing pros and cons and pie-in-the-sky lists of what else I could possibly do with my life. Eric's work was going well; we'd take a hit financially while I figured out what to do next, but I had the privilege of being able to walk away. My life felt out of my control; I needed to let go of something or I'd drown. It wasn't going to be Clementine. It was the right thing to do, but I'd never quit anything this big since leaving physics back in undergrad, and I still remembered how much *that* hurt. This was my entire career. Research is so competitive that once you've got a significant hole in your publication record—from being home with children, for example—you probably aren't going to get back in. There will always be someone ahead of you without that gap.

It felt as though my supervisor had been dangling the possibility of not renewing my contract over my head as a perverse form of motivation. Every time he didn't feel like my results were coming in fast enough, he'd shake his head and say that he perhaps wouldn't be rehiring me. I think he thought I'd rush back to my office and work faster to save my job. What he didn't understand was that I was already at the very limit of what I could manage at that point, and every threat sent me into a panic that was almost paralyzing. I couldn't keep working with an axe over my head. I assumed he was being genuine, so saving him the decision didn't seem like it would be such a big problem. As it played out in my mind, he'd sombrely nod his head, agreeing that I just wasn't cut out for this, and that would be that.

So I was shocked when he was shocked. "But you're doing a great job! You're getting good results! You can't stop *now*," he said. I think my jaw visibly dropped. That he thought I was doing a "great job" was certainly news to me. How big a difference could it have made if he'd let me in on that little secret before I had one foot out the door? If he'd told me I was succeeding at a time when I couldn't see it for myself? I gaped at him, completely unsure what to say. I hadn't come to the decision easily, but once I'd made it, I was committed. It was too late.

When that statement didn't immediately turn me around, it seemed to make him angry.

We had an exchange of words that burned itself into my mind and made it clear to me that he considered me to have been a terrible investment, and that he'd likely think twice the next time he interviewed a pregnant woman.

Those were the last words we ever spoke on the subject, and in the moment, I believed I *had* been a bad investment. My hands shook as I walked quickly back to the safety of my empty office. I think he'd expected tears; maybe a full-on breakdown. Instead, what he got was a polite, stone-faced apology for wasting his time and a refusal to backpedal. This was the legacy of a childhood among grown men. No superior of mine had ever seen me cry, and I'd be damned if he was going to be the first.

I'd never been more ashamed of myself in my life. Less than two years into the career I'd worked toward for more than a decade I was walking away, probably for good. In hindsight, I see that I had other options. But when you're drowning, you panic. It never crossed my mind to try to find a different postdoc because I was sure the problem was me. I never even thought to reach out to Anne and ask for help or advice because I'd always solved my problems on my own for fear of looking weak.

I spent two more months sitting numb in my office, wrapping up the project, carefully detailing everything I'd done so someone else could take over, because that's what we have to do in research or the work will have been wasted. The science was more important than my desire to not be there anymore. I avoided my supervisor whenever I could, not wanting any further discussion of my resignation. But the day I walked out of there for the last time, into the early afternoon sunshine of mid-May, I felt freer than I had in ages. I was a failure, but I could breathe again.

A PAIR OF STUDIES PUBLISHED in early 2020 found that at the undergraduate level, women in the life sciences both outnumbered and outperformed men, but were judged to be less competent. At the upper professional levels, the most significant factor contributing to differences between male and females scientists in both productivity and impact was the high dropout rates of female scientists. Every year that a woman spends in research, she has a 20 percent higher chance of leaving than her male counterparts.[23] According to an eight-year study published in the *Proceedings of the National Academy of Sciences* in 2019, a staggering 43 percent of women leave full-time STEM employment after their first child.[24] These are women who spent a significant portion of their adult lives, as I did, training to be scientists. They wouldn't leave easily, and it's worth asking why so many of them do.

My story wasn't dramatic. My advisor wasn't noteworthy in his hostility or disinterest in being supportive. I hadn't been harassed or abused; I hadn't really even been bullied. He didn't try to ruin my career. There was no *one* dramatic incident that extinguished my desire to be in research. What I'd faced was an environment in which I'd felt under intense pressure to never let anyone see that I had other loyalties in my life. It was a death by a thousand tiny cuts. And that's what makes this story important, because I suspect that's how it is for many of the nearly half of all women in science who leave after becoming mothers. Each time you're made to feel unprofessional for having caregiving responsibilities, each time you're made to feel like a burden for requesting minor accommodation . . . it wears you down a little more. You believe that *you* are the problem. And when the reward at the end of those years of hard work and low pay are far from assured, it doesn't take a PhD to figure out you might be happier and better off elsewhere, no matter how much you loved the actual science and the questions you were trying to answer. No matter how much you wanted to change the world, you still have to make sure your own world is on stable footing. I had wanted so badly for my daughter to look up at her mother and see a scientist. But I also want her to look up and see someone who can walk away from a bad situation.

I tried hard to adapt to fit what the job required of me. But in the end, motherhood changed me more than I could ever have changed myself.

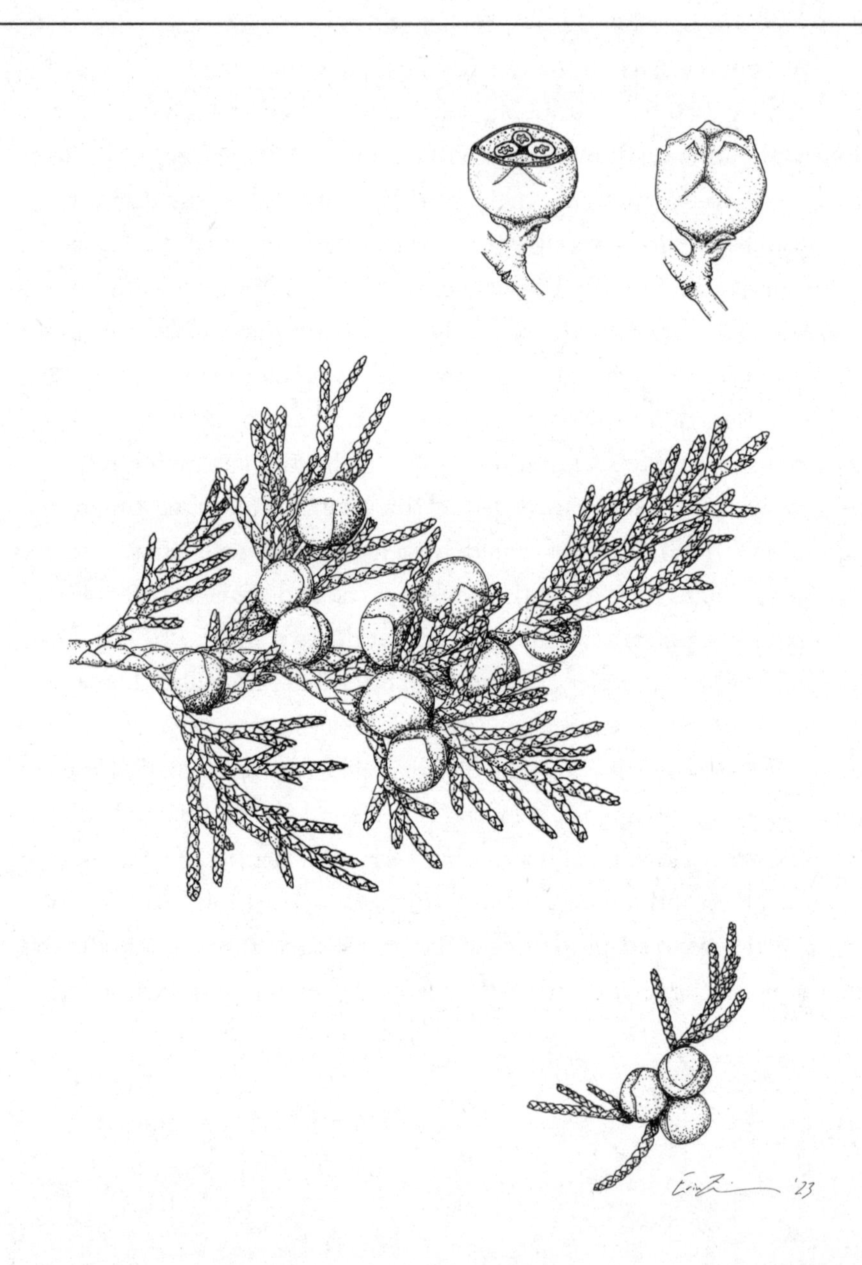

CHAPTER 9

THIS VIEW OF LIFE

> There is grandeur in this view of life, with its several powers, having been originally breathed into a few forms or into one; and that, whilst this planet has gone cycling on according to the fixed law of gravity, from so simple a beginning endless forms most beautiful and most wonderful have been, and are being, evolved.
>
> —CHARLES DARWIN, *On the Origin of Species by Means of Natural Selection*

Back at home again, in the void of unstructured time that comes with unemployment, I could take a breath and consider my options. To my surprise, the silence was less deafening and the limbo less frightening this time than the last. It felt more like an open field of possibility. The sense of relief was intense, but punctuated with waves of grief that struck me at odd times. There were social media posts from friends complaining about lab work or sharing their newest publications, new initiatives to attract girls and women to science, people talking with awe about new species being discovered. Sometimes it was something as minor and silly as being able to pour very precise volumes of liquids with ease when I was cooking, or accurately eyeball a few grams of baking powder, and knowing it was a vestige of a well-developed skill set I'd never need again. In those moments my heart would ache, and I'd try to think about anything other than the life I'd let go.

Clementine went part-time to day care while I worked on figuring my life out, but on our days together, I soaked up the mundane daily routines of motherhood with the appreciation of someone who knows what it is to miss them. For the first time, I savoured her without metastasized exhaustion devouring me from the inside. She wasn't a baby anymore, and wouldn't let me just rock her and look into her eyes like she had when we'd last shared our days, but my happy, active one-year-old offered a lot more possibility for adventure than my baby had. She had some semblance of a schedule and could be trusted to remain awake and in a good mood for several hours at a time while we toured parks, took in music and food festivals, hiked local nature trails, and went out for the odd lunch with friends. I could be happy with her without an abrupt end looming over us; I could start to make up for lost years of understanding the back and forth of mothers and daughters—all the things I'd missed the first time around.

When I thought back to why I chose to be a biologist in the first place, it was because it rewarded looking deeply and really seeing the elegance and complexity of the natural world. I hadn't anticipated then how much I'd enjoy what the writing aspect of science provided: getting to tell others about what you've seen on your adventures. But my hands were always tied by the sort of language that needs to be used in scientific articles—exacting, and with little room for splendour or sentiment. Natural history, with its long history of being intertwined with aesthetics and the appreciation of beauty, granted a little more wiggle room, but ultimately, it all had to be recorded in a modern journal using an impersonal voice. As a writer, I could convey facts without having to divorce them from the beauty and wonder of their surroundings, like

Marianne North painting her plants in the landscapes to which they belonged. And I could let a little of my own voice and thoughts creep in without the self-erasure that's expected of professional scientists.

Could I write about botany for people who *didn't* want dry scientific language? The thought was hard to brush off, but nerve-wracking after what I'd just gone through leaving academia; I couldn't afford to be wrong twice. After all these years, did I have any skills the world *actually wanted*?

It's been a long time since this sort of writing was common in a scientific context, but it wasn't always so. Alexander von Humboldt, who wrote in the early to mid-nineteenth century, was a great example of the blending of beauty and science in writing.[1] He packed his written works full of facts and measurements, but also his own responses to the grandeur of the landscapes around him and what he'd learned from them. He wanted to inform, but also to provoke awe. He wanted science made poetic. Darwin, taking direct inspiration from Humboldt's work, used a lyrical voice in his own writing, meandering among examples and referring to himself and his impressions. This style allowed him to draw in a wider swath of readers and helped his theory gain acceptance among members of the general public more quickly than if he had only written for others in his field.[2]

The bifurcation between lyrical and technical science writing came in the mid-nineteenth century and was tied to both the professionalization of science and the rise of science writing directed at lay readers. Up until the early decades of the 1800s, there *was* no mass reading audience for books about science aimed at amateurs.[3] That began to change for several reasons. "There was a confluence of factors that came together," explains Bernard Lightman, a professor of humanities at York University who specializes in Victorian-era science and women's role in it. "Including advances in printing technology that allowed more books to be published more quickly and cheaply, rising literacy rates, and the establishment of a railway system so you could transport books more easily. And the cost of paper went down, for a whole series of reasons." By the time the 1840s rolled around, science books were being purchased for low prices by a large reading audience. The age of popular science had begun.

In addition to the many books being published, science periodicals were all the rage, and unlike books, they provided a venue for ongoing conversations and contributions from readers. Anyone could write a letter and share an observation they had made, be it while walking in the forest or while looking up at the night sky. At least one periodical went so far as to organize readers into study groups on particular topics and publish reports on the groups' findings in an early nod to citizen science.[4]

The proliferation of science books and periodicals meant that those who could write engagingly about such topics were suddenly in demand. This shift was happening at the same time as botany and other branches of science were professionalizing, when amateur enthusiasts and women in particular found themselves unwelcome in the laboratories and universities into which science had moved.

"They were prevented from becoming a part of the scientific community," says Lightman. "Women weren't allowed into universities, could not get university degrees, therefore could not get scientific positions. So there were only a limited range of opportunities for women in the nineteenth century who wanted to be involved in science. But there was nothing to prevent them from writing books. And that's where a lot of women found their place in the scientific world." Not only had they found their place, but the female science writers of the time were granted a level of authority for their knowledge that they'd never been able to have in scientific research itself, and that perceived authority increased as the century wore on and the expectation that they write primarily for children faded. As a group, these women were productive, influential, and widely read.[5]

As scientists moved more toward an impersonal and highly technical form of writing that allowed them to communicate efficiently among themselves while excluding outsiders, popular science writing was in flux. Writers—particularly the women, Lightman notes—were experimenting with various narrative styles and devices borrowed from fiction that allowed them to convey facts in an accessible and entertaining way.

A prime example of this is Arabella Buckley, a British popular science writer born in 1840 and, like Lydia Becker, a female correspondent of Darwin. Buckley wrote primarily for children but experimented with different narrative devices for doing so. In one of her books, *The Fairy-Land of Science*, she wrote fairy stories in which the fairies are actually the invisible forces of nature, such that for children, it was a storybook, but for the adults reading the book aloud, the science was presented as complex metaphors for them to consider. [6] "Buckley understood that when you're writing for children, you're actually writing for two audiences—the adults who are reading the stories to the children, and the children," says Lightman. "It's actually very sophisticated, that book."

Buckley also experimented with a style known as the evolutionary epic. This type of narrative goes back in time to explain, from the beginning, how a living thing familiar to us today came to be. Buckley wrote two such books, on invertebrates and vertebrates, showing how the groups developed over time, changing and adapting in order to survive. In another example of an evolutionary epic, an essay called "The Origin of Walnuts" by John George Wood, the author explicitly sets out to show that the evolutionary journey that produced something as humble as walnuts can be as awe inducing as the one that produced humans. "The whole idea was that you could find things that people have encountered in their everyday lives," Lightman explains. "You didn't have to be working in a laboratory. It's all there, it's all in front of your eyes. Nature surrounds us."

Other writers of the time explored still other styles of imparting scientific information, including structuring a natural history book as an exploration of one season at a time, and writing stories that were outright fictional but heavily footnoted with scientific explanations of the phenomena being discussed in the story. This point in history was a fork in the road when scientists and science writers set off in different directions. Professional scientists worked to exclude as many people from their self-made ivory tower as they could in order to reinforce their authority, while science writers strove to find new ways to *include* as many people as they could, using language that was accessible to any literate adult and basing their stories in familiar scenes and objects.

"The popularizers have a very different notion of epistemology and how knowledge is discovered. For them, anyone can be a scientist, anyone can study nature. You don't need this expert training in a laboratory," says Lightman. "They quite explicitly say, 'You don't have to go on the voyage of the *Beagle*, you don't have to go halfway around the world to make scientific discoveries. You can look in your own backyard. And if you observe nature carefully, with a devout mind, looking for design, you will make important scientific discoveries.' So there's this kind of democratic notion that the study of nature is open to all. And this gets a lot of people interested in science."

Perhaps just as important, the writers created space for the experience of wonder. Readers were meant to feel that they were, as Lightman writes, "walking beside the naturalist on an adventurous quest for knowledge."[7] They sought out the unique elements that made the world feel magical. What's more, the observer's emotional response to what they saw was considered worthy of attention.

Like my lost childhood afternoons of picking at pieces of straw and peering into flowers, trying to see the hidden things, the act of noticing was encouraged and valued. Curiosity and fascination weren't things to relinquish in the face of adulthood. Writers told the curious that they could be part of the wonder of the natural world if they just took the time to look closely enough.

To me, this meant that at a moment when I felt excluded myself, I could choose to be an active part of including others. Years of conditioning that led to my looking at the world in terms of scientists and "everyone else" would need to be left behind, but understanding the history helped me to see that it was an artificial line that had, after all, been set up to exclude people like me anyhow. Now that I knew what that felt like, I understood why the tent needed to get bigger.

I BEGAN FREELANCING AND SPENT the next couple of years careening from one topic to the next, enjoying the freedom but failing to hit on some winning combination of interesting and reasonably well-paying. I was accustomed to feeling underpaid, but freelance writing opened my eyes to new depths of being undervalued for one's expertise. What paid the best, as it turned out, was medical ghostwriting. A doctor wishing to publish an article based on their research or a talk they had given, but too busy to do so, would pay me to take the information they provided and write an article that would bear their name. It was interesting learning about all kinds of medical conditions and treatments and putting together educational articles about them, but my name never appeared on my work, and in some cases, I wasn't even allowed to say that I'd done it. I wasn't building my portfolio or creating anything I could point to and be proud of. I thought of all the people I knew through my years in research and how, to them, it must appear that I had simply had a child and dropped off the face of the Earth. As though I was no longer interested in a career in science. I hated that some would undoubtedly take this as confirmation of their bias against training and hiring women.

More and more, the people I knew in science faded from my life and were replaced by people who knew me only as Clementine's mother. I no longer had coworkers, after all. To the people I met in the community in those days just before the pandemic made working from home common, I was a stay-at-home parent. What I did for a living was invisible, so no one asked. During my postdoc, I'd needed to hide being a mother and present myself each day only as a researcher, but now the pendulum had swung to the opposite extreme, and I was only a mother. I didn't like it, not because it isn't a valid identity, but because I didn't want it to be the entirety of mine. I'd spent most of my life not wanting children, only to have motherhood consume me. I've noticed that very often when people meet you as a mother, that is all they will ever think of you as—whoever you were before, with your hobbies and dreams and idiosyncrasies . . . it doesn't matter anymore. You are a local mother of *x* number of children. The higher *x* is, the less you are presumed to have accomplished.

Friends and family members were largely silent on my career. No one knew what, if anything, I did, so no one brought it up. Instead they asked about Clementine—the one facet of my life they understood and were sure still existed. When I occasionally bumped into former coworkers from my postdoc, they asked how my little one was doing, and when I was going to have another. They never asked what else I was doing. And if they had, what did I have to show? Who even was I now?

I thought I'd coped with most of my feelings about leaving science by this point, but as I realized how the world now saw me, a lot of shame and anger began to bubble back up. I started to have moments of panic when I thought about my CV and how, if I was quick, maybe there was still time to find another postdoc and go back to doing research before the gap in my publication record became definitively career ending. An unnerving voice in my head kept insisting, *There's still time . . . go back . . . go back!* It took all my self-control at these moments not to act on that impulse.

I thought often about how lucky I'd been to get to travel as much as I did while I was in research. All the things I'd gotten to see and hear and feel that most people don't. Daybreak in the jungle pierced by the sound of howler monkeys as I hung half awake in my hammock. Orchids bathed in the spray of a waterfall, perhaps never seen by anyone outside that place. Centuries-old specimens of plants that no one will ever see alive again. Lines of genetic code telling me something completely new to human minds. All those places and sights and moments of serendipity—that feeling of being part of the wider world—seemed closed to me now. My life was lived at home, and my sphere of influence felt smaller than it had been in a very long time.

When I looked for inspiration for the life I found myself in, I kept coming back to Charles Darwin. Not because he was a brilliant scientist or a famous historical figure, *per se*, but due to the circumstance of that fame and brilliance. We associate Darwin with his voyages aboard the *Beagle*, which he undertook when he was only twenty-two. But the theory of natural selection came much later. For as much as we may think of Darwin as an intrepid man of science, his time aboard the *Beagle* was the only trip abroad he ever made. For five eventful years in his early adulthood, he saw the world. And then he never left Britain again. His poor health following his early voyage wouldn't allow him further travels, and country living was better for his well-being than the city. He settled in the Kentish countryside, in a secluded village of fewer than five hundred people, and became a father many times over.

Darwin was constitutionally well-suited to a quiet life in the country.[8] Though he had many friends in high places in London, he thoroughly involved himself in the life of his village, getting to know his neighbours, acting as a local magistrate, and even listing himself as a farmer in a local directory. He helped raise his eight surviving children—two more died in very early childhood—and was by all accounts a loving and devoted father. He enjoyed the routines of his life and spent a great deal of time in his garden, avoiding the limelight of London scientific society as much as possible. More than anything, he seemed to want to be alone with his ideas.

How does someone so modest and retiring, so entrenched in rural life, go on to change the world? In short, he made the most of his surroundings and the reach available to him through correspondence. Unlike the metropolitan men trying to professionalize the life sciences, Darwin never worked in a laboratory—his home and gardens were the sites of his experimental and observational work. As Janet Browne writes in her biography, *Charles Darwin: The Power of Place*, "Over the years, Darwin bred pigeons, grew pots of seeds in his outhouses, observed bees moving across his flowerbeds, tracked worms in the fields that he saw from his drawing room window, counted blades of grass in his lawn, watched his infant children in the nursery, and pondered the twists and turns of climbing weeds in his hedges, all the while seeking the detailed evidence of adaptation in living beings that he believed to be the keystone of his project."

He also used his connections in the scientific world to pull in information on whatever he needed to know. But it wasn't only other scientists he tapped for answers. "He also hunted down anyone who could help him on specific issues, from civil servants, army officers, diplomats, fur-trappers, horse-breeders, society ladies, Welsh hill-farmers, zookeepers, pigeon-fanciers, gardeners, asylum owners, and kennel hands, through to his own elderly aunts or energetic nieces and nephews," writes Browne. With his correspondence, Darwin took advantage of the increasing European presence abroad and the resulting improved postal service, sending messages across Asia, Oceania, the Caribbean, North America, and the Pacific Islands. Over 14,000 letters are known to have been sent or received by Darwin, with many more presumed lost.

The way he worked looked nothing like the professional men of science of that time. Yet he met with incredible success.

"That's the ultimate irony of this whole interest in Darwin," says Lightman. "Darwin became a poster boy for modern professionalized biology. But if you go back and you look, he was anything but a professional scientist, even though he wanted to be considered one. He didn't have that kind of training, and he never got a scientific degree. His training was on the spot, it was while he was out in the field, while he was on the *Beagle* voyage. He did not ever work in a laboratory. He had his own makeshift laboratory on his estate. So Darwin was not a professional scientist at all. He didn't even identify himself as a biologist, he still thought of himself as a natural historian. When you look in *The Origin*, you won't find the term biology in there, you'll find the term natural history. In a sense, he is the worst possible example contemporary scientists could choose as an emblem for modern biology."

Darwin made his home the informational hub of a vast network that supplemented what he could do with his own hands in his own garden and workspaces. He reached out into the world, drew in facts and figures, and used them to create something new and extraordinary. What's so important about this, to me, is that one of the most groundbreaking scientists of all time did the key work of his life alone in his home in the country. The circumstances of his life made adventure, and even a metropolitan existence, impossible for him, but he did the work anyway. Quietly, and without eyes upon him.

Lydia Becker looked at Darwin and saw the same thing I did. When speaking to the similarly homebound women of the scientific society she formed in Manchester after being rejected from local men's societies, she pointed to Darwin as a role model for what can be accomplished with careful observation and diligent study, saying, "Such an example should encourage us to go and do likewise." She championed the idea that anyone could contribute, telling her audience, "Any one of us might begin a series of patient observations . . . if the observations were carefully and accurately recorded . . . the result would be something of real, if not of great, scientific value."[9]

Like Becker, I can't help but feel disappointment toward Darwin for the extent to which his normally forward-thinking nature failed him when considering the intelligence and abilities of people who didn't look like him. But also like Becker, I've felt comforted by the magnitude of what he was able to accomplish while tied to his home. I think many of us—primary parents in particular—feel that life clips our wings at a certain point. The window for adventure closes, and we find ourselves wondering what to do with the dreams that remain. But if a chronically ill father of eight living in nineteenth century rural England can shake the very foundation of what we know about life itself, surely those of us tied to our homes and responsibilities, yet with the technological power of the twenty-first century at our fingertips, can still harbour hopes of making a difference.

IN THE SPRING OF 2019, I once again found myself at Kew Gardens. It had been three years since I left research, almost two and a half since I started writing about science professionally, and just over a year since my second daughter, Juniper, was born. I was happy with my young family but still adrift, uncertain where my career was headed. Eric and I hadn't had a vacation since we'd finished school, so we brought our daughters to the UK for a week. I planned a day at Kew for us so I could show them one of my favourite places in the world, a place I called home once in another life. Any seasoned parent knows that it's folly to set your hopes on your children loving something simply because you do, but that day, I couldn't help myself.

We walked the foliage-framed pathways on that warm, bright day in May as the whole of the gardens seemed to burst into bloom. My favourite place in my favourite season . . . I expected big smiles and happy memories. Instead, I got crowds and grumpy children. An hour or so into our saunter about the grounds, I felt worn down from a crying baby and an irritable preschooler. I tried to show Clementine around in what I considered to be the best of the greenhouses, the Temperate House—the world's largest Victorian-era glasshouse. The sparkling glass panes and whitewashed wrought iron were a vision in the morning sunlight. The smell of flowers and warm foliage was intoxicating. I wanted to tell Clementine about the happy times I spent here as a visiting scientist, and how I'd loved this place. As I crouched down and put my arm around her to take a picture of the two of us with the greenery behind, she refused to smile because she was bored and wanted to go back to the hotel and watch cartoons. On a different day, I might have found her stubbornness endearing. I sighed, deflated, and walked her back outside. Eric saw my expression as we emerged and offered to take my picture in front of this place I'd been so eager to return to. I tried my best to smile. When he handed my phone back to me and I looked at the picture, all I could see was a tired, sad woman who looked just like the frustrated, worn-down mothers that used to make me not want to have children.

Of course, we all feel this way sometimes as parents, because parenting is hard and children have bad days, not because we have bad children. But in the moment, I was struck with the full force of how far I'd fallen in my own estimation: from walking into Kew as a promising PhD student with her career in front of her and big plans for herself in 2010, to a tired, washed-up, never-was who had to push along with the crowds because no one knew her anymore, nine years later. I no longer got to see the hidden places the tourists couldn't go or take tea with scientists and talk about new discoveries.

I asked Eric to take the kids for a while, and as they walked off down the path, I sat down on the steps of the glasshouse and buried my face in my hands. I thought back to Larry's advice. This was where I once believed I could learn to be the best. But now I was just another woman who amounted to nothing more than having kids and disappearing.

My dark night of the soul, on a sunny day, surrounded by beauty. I walked away.

As you do in a tourist attraction, I ended up in the gift shop. It's not where you expect to find your moment of truth, but wandering around there, I found myself in front of half a dozen large wooden tables piled high with books about plants. They were nearly all written not by scientists, but by people who just loved plants, knew a lot about them, and wanted to share that.

It was a busy day, and I watched people milling about among the tables, many with several titles clutched in their arms. I thought of the female science writers of the nineteenth century, using their craft as a way of digging their heels into the worlds they'd been pushed out of, as a way of being involved on their own terms. For nearly two centuries, women who loved botany but found themselves on the outside of the ivory tower had turned to pens, typewriters, and keyboards. They did it to teach, to revel in what they'd learned and fallen in love with, and to help others see the wonders all around us that we often fail to notice. More recently, they started doing it to sound the alarm about the destruction and devastation we're bringing to our own world and the species we share it with.

Sometimes you just need a glimpse of where you're headed to find the drive to keep going.

As biologists, we try to press order—hard, clear dividing lines—on a world of fuzzy boundaries. We try to find a story that makes sense and is supported by the evidence we have available in order to explain the world to ourselves. We all do this in our own lives as well, so our personal histories can have the clarity and logic that a narrative allows. In science, of course, we always have to be ready to abandon a story or an explanation if the weight of evidence shifts away from it. That can be hard. It's even harder in our personal lives, rooting out the stories we're telling ourselves that no longer fit. And sometimes the letting go is a process.

Leading up to the birth of our second child, Eric and I left the city once again, and for the last time. I've come full circle, back in a farming community of southern Ontario, not far from the field-lined, dusty road that I grew up on—a place where, after all this searching, I can finally put down roots. My stacks of scientific papers and lab books are still packed away in boxes, though they no longer feel like an itch in the back of my mind, calling me to keep striving for something I no longer want.

I'm trying hard to teach my children to notice the details and to look deeply at the world, hoping that close attention leads them to fall in love just as I did. With my daughters in particular, I'll accompany my message of chasing their dreams with the reminder that what they give up for those dreams is for them alone to choose, and that dreams can change with time. Looking back, I'm immensely thankful to the exhausted, nerve-bare, and almost broken young mother that I was for seeing that what I was doing wasn't worth the cost. It's been seven years now since I abandoned the career I'd dreamed of and worked toward for so long. I'm hardly alone in my story; many others have walked away from careers they strived toward for similar reasons. Though it isn't easy, all of us have to find another place we can be happy and contribute. But there are so many ways to make a difference.

CHAPTER 10

BOTANIST AT LARGE

> The one process ongoing [now] that will take millions of years to correct is the loss of genetic and species diversity by the destruction of natural habitats. This is the folly our descendants are least likely to forgive us.
>
> —EDWARD O. WILSON, *Nature Revealed: Selected Writings, 1949–2006*

It's a Thursday afternoon in April 2021, still the height of what I'll remember as the "everything is virtual now" era that accompanied the COVID-19 pandemic. I'm sitting at my desk in front of my laptop, which is showing the faces of five other women, each sitting comfortably in a bright kitchen or book-lined home office. They range in age from thirties to eighties and are spread across the United States. We're all here to learn how to help digitize herbarium specimens as volunteers, part of an online event put on by WeDigBio[1] and the New York Botanical Garden to teach the public about collections, herbaria, and the importance of digitized specimens. These women and I are part of a larger audience that's been split off into smaller groups to learn the ins and outs of the task at hand.

There are two or three men among the two dozen or so people that have turned out for the event, but it is overwhelmingly female and mostly over fifty. Not what you'd call the stereotypical face of science.

After we've had a few minutes to introduce ourselves and make small talk, another volunteer, this one with many hours of experience under her belt, enters the chat room to begin our instruction. We're shown a high-resolution scan of a herbarium voucher collected in the late nineteenth century. The information contained on the specimen's label, spread across it in elegant sloping script, needs to be deciphered and entered into a database in a standardized format to make it searchable to researchers who can then examine the digitized image or request a loan of the actual specimen. Much of the task is making sure that location data is properly recorded, so we learn about resources for finding the correct names of municipalities and administrative districts of far-off places. We learn about collector numbers and how to properly list expedition names and secondary collectors. Hardly exciting, but there's a certain charm to looking through these tiny windows to distant places and times and picturing them in your mind, aided by the short descriptions of the plants' surroundings given on the labels. And when you've completed the transcription, a counter on your personal volunteer page goes up, tallying your lifetime contribution to the cause. Using a website called DigiVol,[2] transcription tasks are organized into "virtual expeditions" grouped around particular regions or groups of species, further playing up the feeling of being on an imaginary adventure.

This is one of the many faces of citizen science in the 2020s: helping out on a nineteenth century botanical expedition to Tanzania from your kitchen table.

The term citizen science is relatively new, having been coined in the 1990s,[3] but the concept is very, very old. Prior to the professionalization of western science in the nineteenth century, nearly *all* science was citizen science, in a sense. Some naturalists were sent out by their governments in an official capacity, but the vast majority of the people we think of as the Great Naturalists did what they did because they were curious and wanted to share what they'd found. Well-known explorer-naturalists like Humboldt[4] and Alfred Russel Wallace[5] went out on adventures they conceived of and financed themselves that ended up building our understanding of the world, but so too did countless unknown contributors who made less flashy contributions that also comprise the foundation of our knowledge today.

One impressive example is data on the flowering times of cherry trees in Kyoto.[6] Using hundreds of diaries and chronicles spanning over a thousand years, researchers were able to assemble a nearly continuous record reaching back to the ninth century. The flowering of cherry trees between late March and mid-April in western Japan depends closely on general temperature conditions during flower bud development, which occurs in February and March. The flowering dates therefore allowed the reconstruction of the mean March temperature each year. This data in turn revealed repeated cold periods that corresponded to the long-term solar cycle, which has a period of between 150 and 250 years, and suggested that variations in local temperature brought on by fluctuations in the solar cycle lag fifteen years behind the solar cycle itself. All this from long-forgotten locals dutifully noting the cherry blossoms over the course of a millennium.

Today's hard line between experts and knowledgeable amateurs, though it may feel immutable, is a fairly recent development.

Modern citizen science[7] projects can take many different forms, but at the most basic level, they are any task completable by a nonexpert in direct service of answering research-based questions. Sometimes this means gathering and organizing data, while in other cases it can mean aiding in the analysis of that data. In addition to transcription work, there are projects with tasks as diverse as recording sound levels of noise pollution,[8] noting the progression of invasive pests,[9] and even gamified DNA sequence alignment[10] and protein folding.[11] The most common type of citizen science project in the life sciences involves biodiversity monitoring.[12] These projects note the presence (and occasionally absence—an undervalued type of data point) of certain species in certain places, as well as useful auxiliary information such as flowering times, pest damage, or pollinator species. Data is often recorded on volunteers' smartphones but can also be entered on data sheets or in field notebooks according to the needs of the project. Inviting citizen scientists to contribute data can vastly increase the scale of a project versus what researchers can do on their own with a limited amount of time and funding.

The Global Biodiversity Information Facility (GBIF),[13] an international data network that provides open access to biodiversity data, estimated in 2018 that half of the records it maintains—numbering over two billion as of early 2023—have come from amateurs, and that their records have been included in over 2,500 peer-reviewed scientific papers.[14] The digital observation platform iNaturalist, as of 2021, had racked up fifty-six million observations, with that number roughly doubling annually.[15] (These two numbers are not entirely independent, as iNaturalist observations can be and regularly are included in GBIF.[16]) Another online platform, Zooniverse, can receive millions of observations daily[17] and has contributed to over sixty scientific publications based on biodiversity data alone, with many more related to the physical sciences and humanities.[18]

The data collected by volunteers working on biodiversity conservation projects allows researchers to form a picture of the abundance, distribution, and habitats of one or a group of species, and to keep tabs on any ongoing changes in these elements due to factors such as climate shifts or human disturbance.[19] Carefully designed projects can replicate historical data gathering, allowing direct comparison across long periods of time and making recent shifts more starkly visible. With sufficient instruction and explanation, the collected data can be at or near the quality level of that collected by the researchers themselves and can be used to expand or fill gaps in the scientists' own data. This data and subsequent analysis can then be used to inform policy changes at local, regional, and national levels.

The benefits to researchers of deploying citizen scientists to gather data are fairly evident: more data gathered, or in some cases analyzed, at lower cost than if they had done it all themselves. A 2015 estimate put global citizen science participation at 1.3 million volunteers across nearly four hundred projects, contributing a value of $2.5 billion USD. "For biodiversity science," the paper's authors declared, "the era of ivory tower science is over."[20]

With proper project design and execution, citizen science initiatives can be immensely valuable. But it's worth asking what the volunteers themselves get out of it. If initiatives are to retain enough people to achieve research goals, it's important to provide value to them based on their personal motivations for participating. The answer to this depends to some extent on the nature of the project. In more traditional projects, people met up in person and received training, such that they benefitted both through education and through the social aspect of meeting people with similar interests. Today, when many (though not all) projects are done remotely, primarily individually, and without significant or ongoing training, the answers become more varied.[21] Even under this new paradigm, volunteers gain awareness and knowledge of living species and the environment and are provided with extra motivation to get out into nature or get moving. For some, there's a competitive element to seeking out rare species, or a "*gotta catch 'em all*" impulse when adding to their personal species lists, an impulse long familiar to avid birders. And for some, just knowing that they're contributing to science and are part of something larger than themselves is the main motivator.[22]

Mobile technology in the form of apps created by platforms such as eBird, Nature's Notebook, Zooniverse, and iNaturalist allows anyone with a smartphone to upload observations that are automatically GPS localized. This has enabled even very casual users to be involved in a low-commitment way, leading to the term "cyberscientist" for volunteers doing citizen science solely via the internet.[23] Mobile-based platforms are a good way to attract younger volunteers, not only because of their comfort with the technology but because those platforms require less time commitment from a segment of the population that, unlike many older volunteers, doesn't necessarily have a great deal of time to offer.

The ubiquity of phones and the possibility of using them to collect data with minimal instruction has aided with the gamification of certain tasks, couching them in a fun, competitive user interface where volunteers can do anything from building models of brain cells to shooting at parasites on images of malaria-infected blood.[24] In a somewhat surreal inversion of games being integrated into citizen science projects, the 2019 first-person shooter game *Borderlands 3* actually has a citizen science game integrated within *it*; in the game, you can walk up to an arcade machine in an infirmary and play *Borderlands Science,* a real-world project helping to map the human gut microbiome.[25] In an age of mobile games available to fill every minute of down time, this approach—gaming with a purpose—can be very motivating for some users.

A facet of citizen science that is increasingly relevant to project design is the use of artificial intelligence, and specifically machine learning.[26] As we saw earlier, this approach, coupled with computer vision, has a great deal of promise in streamlining the process of species identification, allowing the limited time of the limited number of professional taxonomists to be better spent with truly hard-to-pin-down specimen identifications. This same idea is in use in mobile species observation and reporting apps, allowing the program to give targeted (though not always accurate) suggestions as to the species shown in an image based on its appearance and location. Early integration of computer vision into iNaturalist became a key means of quickly sorting observations and has come to be used in nearly half of all initial identifications on the platform, increasing the likelihood of an observation reaching research-grade quality.[27] Computer hearing can also be used to distinguish among sounds such as birdsong that volunteers may record in the field. These same functions can be used to flag user identifications that may be in error and need further verification.

In either case, computer-generated suggestions for species ID create a friendly and less intimidating user interface for those new to taxonomy and allow a feedback loop whereby high-quality observations can then be fed back into the algorithm to better train it to generate IDs.

As the machine learning algorithms in use continue to improve, they have the potential to decrease both the difficult and the more mundane, boring tasks needing to be done by humans, including high-level classification and sorting of species, transcription of handwriting, the filtering out of low-quality observations, and the organization of captured information. There is a risk with AI, however, of overestimating the computer's abilities and letting the machine have the final say, when in fact human experts know better. As NYBG's Damon Little pointed out when we spoke, it's difficult to get a computer to tell you it doesn't "know" something. Checks and balances need to be built-in to avoid handing over too much authority.

Including contributions from citizen scientists can vastly increase the amount of data available to a biodiversity monitoring project, and can even broaden the physical area that it covers beyond what would otherwise have been possible. But with that huge advantage come a number of potential drawbacks that must be acknowledged and mitigated in order for citizen science data to be top-notch. Casual participation in the absence of any training by experts in the field can make data quality a problem. High participation rates help drown out the effect of these, as does the weighting of users' input differently based on the quality of their past observations. In the case of iNaturalist, an observation isn't considered research-grade until two independent users agree on a species-level identification. But the user base of the platform is now large enough that the median time it takes for this to happen is only *four hours*, and close to 60 percent of all observations will reach this ranking.[28] Crowdsourcing not just observations, but identifications, has helped enormously.

Biases of citizen science data tend to be of known kinds that can be at least partially mitigated with artificial intelligence and clever analytical techniques. For example, statistical methods used in analysis can help take into account the geographical "unevenness" of the observations of volunteers who tend to stay closer to paths and inhabited areas, or their tendency to ignore common species while focussing on sightings of rarer ones, thereby skewing the recorded abundances of both.[29] Volunteer-contributed data will always need the careful oversight and verification of experts, but with careful attention and curation, it can approach the quality of that generated by the researchers themselves.[30]

Data generated by citizen science initiatives has been used to strengthen scientific research, guide local environmental actions, and influence national-level policy. It adds greater weight of evidence to databases that may have future uses we've not even conceived of yet. It can be difficult to draw a direct line from collected data to policy changes in many cases due to the way it is subsumed within scientific publications. However, to give one example, data from the North American Breeding Bird Survey has been used to inform the Endangered Species Act, the Migratory Bird Treaty Act, and the National Environmental Policy Act in targeted protection of certain bird species and their habitats.[31]

For the people who take part, this sort of work can meet a need to be involved in something important and beneficial to the greater good, even if the volunteers themselves have only a little time to give. Like the science writers of past centuries trying to include as broad a swath of people as they could, citizen science is open to those of almost any age, skill set, educational background, and level of income. It asks only a bit of time, a genuine interest, and a willingness to notice.

THE BIODIVERSITY[32] CRISIS BROUGHT ON by the actions of humans—in particular agriculture, resource consumption, and climate change—is growing ever more dire. Almost six hundred seed plant species are known to have become extinct since the eighteenth century, though this is likely an underestimate due to the high probability of dark extinctions. Already it is believed to represent a pace as much as five hundred times greater than the background extinction rate for plants. Multiple estimates have declared approximately 40 percent of all vascular plants to currently be at significant risk of extinction, with levels much higher for particularly vulnerable groups like cycads, which sit just below 70 percent.[33] Startling enough on their own, but these numbers hide a greater complexity to the situation.

When the area of a certain habitat—a savanna or a marsh, say—decreases or is substantially altered, the number of species it can support also decreases in a defined way known as the species-area relationship. Extinctions will then occur until a new equilibrium is reached. In organisms such as plants, with long lifespans and dormant stages in the form of seeds, this return to equilibrium can take a long time. This is called extinction debt—the habitats are already lost, but the toll hasn't yet been paid. Like light from a distant star that's already gone nova, we can't see the fallout yet, even as climate change is predicted to cause further reductions of up to 50 percent in species' potential range sizes during this century. Given the changes in land use since the 1700s, the circa six hundred extinctions we've seen is much lower than the number expected, suggesting there are many more already set in motion than we may fully realize. But as long as those species living on borrowed time are still here, something could be done to save them. Assuming we can recognize them in time, they could be targeted for special conservation action, moved to a suitable habitat within their historical range, conserved in living collections, or potentially be seed banked. But this all depends on our ability to recognize them.

Massive geographical and taxonomic holes remain in global assessments of risk for plants and fungi. Even the estimated 10 percent of plants assessed for the Red List is deceptively uneven, with only around 6 percent of ferns and 1 percent or less of mosses and algae included within that count. And these numbers are for species that are already known to science. For the many, *many* that are not, high threat levels are expected due to their often having small range sizes. Geographically, both temperate and tropical Asia, as well as parts of South America, among others, are underrepresented in assessments. Filling these gaps in assessment coverage is crucial because the lack of information makes well-targeted conservation action for those regions and species impossible; unassessed species are essentially invisible to funding agencies.[34]

In a curious, motivated public lies an incredible resource to help combat, or at least understand, that threat so that further action can be taken. We now have the technology and digital infrastructure to mobilize volunteers worldwide and gather data at an unprecedented rate.

Of course, even with a motivated public involved, that effort needs to be accompanied by the guidance and analysis of experts. This is critical work, and participants on every level must have the support they need to get it done. We mustn't fall into the trap of thinking that more volunteers will require fewer scientists, thereby allowing even more expertise in natural history research to slip away.

If the efforts of citizen scientists show anything, it's that awareness can spur action, even on the level of government policy. And it's my hope that awareness of this particular problem can spur action as well, because nothing short of significant policy change on the part of scientific funding bodies will halt a loss of expertise that will be far more difficult to restore once it's gone. Programs and funding need to be in place to ensure that skills and knowledge are passed from older to younger researchers, and that taxonomic work, collections, and long-term observational research are not done at the expense of one's career advancement.

As I write this, my one-year-old twins, my last babies, are in the next room quietly destroying my house, as toddlers do. This manuscript has come into existence alongside their first breaths, first steps, and first words. Children, even arriving two at a time, do not stop motivated women from chasing their goals. Mothers can achieve whatever they set their minds to, given sufficient support. But there has to be a good reason for it . . . a reasonable hope of reward at the end of it all. The current state of research jobs is such that, for many women, there isn't. Staring down the barrel of a decade of short-term contract appointments at the same time you need to be starting your family would and does make even the most dedicated scientist think twice.

The common framing of scientific research as a calling rather than a job leads people to accept a lot of things that they shouldn't, like unreasonably long hours, low pay, and long-term instability. Science is being called upon to address ever more serious problems, but part of supporting science is supporting the people who do it.

The natural world as we've known it is in serious danger, as are we as a species. I believe that collectively we possess the skill, intelligence, and compassion to solve our problems with science, both directly, to generate solutions, and indirectly, to influence policy. But we need to be less short-sighted and easily distracted than we are. In a fast-paced age, we need to slow down, notice, consider, and contribute what we can.

As the volunteer transcription event draws to a close, a couple of the more experienced volunteers recall some of their more interesting finds during the work—specimens that were particularly old or unusual, or that came from some very remote locale. A woman in her sixties comments that she doesn't travel and would certainly never have seen these plants any other way. I felt a twinge, knowing that this would likely be true for me now as well. But to look at it another way, what a gift that those on the outside of the ivory tower can see these hidden treasures now. We are Darwin in his countryside home, poring over life's mysteries, but without the need of status or connections. We live in a time when anyone who wants to help can.

I was told as I began my career that in science, there's always room for the best. But done right, there's room for everyone. There has to be.

ACKNOWLEDGMENTS

First and foremost, thank you to the two people without whom I would be lost—Eric Chevalier and Caitlin MacGregor. Cait: life feels so much kinder when you know there's someone who's always on your side. You are my first reader, my cheerleader, my best friend now and always. I don't know what I'd do without you. Eric: thank you for everything you do to make this beautiful life we're living possible. You've always believed in me, and you never give up. Je t'aime.

Thank you to the botanists who shaped my time in research: Anne Bruneau, for giving ambitious women great opportunities; the members of Labo Bruneau; Larry Peterson, for giving me my start in science and a vision of how things should be. To the late Usher Posluszny, and the members of the Peterson-Posluszny lab. And of course, to Karen Redden, my model of a fearless adventurer.

To the people who permitted me to write about them and their families, and made my life richer for their presence, however brief: Lyn Baldwin and her daughter Maggie; Pat, Donna, and Sarah Herendeen. You were all so generous with me. To Carole Sinou, for being my example of adaptive motherhood, even if I didn't realize it for a long time. You are just as impressive to me now as you have always been.

Thank you to all the scientists and historians who were kind enough to speak with me for this book: Barbara Thiers, Pamela Diggle, Pamela Soltis, Alice Tangerini, Damon Little, Renate Ysseldyk, Ann Shteir, Bernard Lightman, and Spencer Barrett. And to the Natural Sciences and Engineering Research Council of Canada for funding my research as detailed in these pages.

Thank you to Kirsten Reach, Mike Lindgren, Sofia Demopolos, and Athena Bryan, and to my agent, Jessica Papin, for your patience with this first-timer, and for helping me shape this mass of ideas into a coherent story. Thanks to the whole team at Melville House for getting this book out into the world. A big nod to Ashleigh Renard for gifting me its title, and to those who took the time to read and praise *Unrooted* ahead of its publication.

Thank you to each and every person who cared for my children during the time I wrote this book, and in particular Amy Stewart and my local daycare. Quality childcare is rarely afforded the appreciation it deserves. To the Bostwick branch of the London Public Library, and the Wild Goat Café in Rodney, Ontario, thanks for the quiet space to think and write.

Thank you to Les and Nancy Zimmerman, for all the listening and commiserating. And to Ashleigh Downing, for being a constant friend. Thanks, too, to the fellow writers whose support means so much: Sarah Boon, Melissa Sevigny, Antonia Malchik, Kim Steutermann Rogers, Kimberly Moynahan, Elizabeth Hilborn, and many others whose names I'm not able to include here but who have made my life easier with their support and kindness.

And finally, thank you to my children, for each being your unique, colourful selves. You make this all worthwhile.

FURTHER READING

A by-no-means-exhaustive list of books on some of the main topics covered here.

On Plant Taxonomy, Structure, & Collections

Herbarium: The Quest to Preserve and Classify the World's Plants, by Barbara M. Thiers

Plant Families: A Guide for Gardeners and Botanists, by Ross Bayton & Simon Maughan

The Art of Naming, by Michael Ohl

Flowers: How They Changed the World, by William C. Burger

The Lost Species: Great Expeditions in the Collections of Natural History Museums, by Christopher Kemp

On Drawing Plants

How to Draw Plants: The Techniques of Botanical Illustration, by Keith West

Botanical Sketchbook, by Mary Ann Scott & Margaret Stevens

The Laws Guide to Nature Drawing and Journaling, by John Muir Laws
Field Notes on Science & Nature, edited by Michael R. Canfield

On the History of Botany & Natural History

Finding Order in Nature: The Naturalist Tradition from Linnaeus to E.O. Wilson, by Paul Lawrence Farber
Imperial Nature: Joseph Hooker and the Practices of Victorian Science, by Jim Endersby
The Great Naturalists, edited by Robert Huxley
Sex, Botany, & Empire: The Story of Carl Linnaeus and Joseph Banks, by Patricia Fara
Charles Darwin: The Power of Place, by Janet Browne
The Reluctant Mr. Darwin: An Intimate Portrait of Charles Darwin and the Making of His Theory of Evolution, by David Quammen
The Flower of Empire: An Amazonian Water Lily, the Quest to Make it Bloom, and the World it Created, by Tatiana Holway
Flower Hunters: Adventurous Botanists and the Lasting Impact of Their Discoveries, by Mary Gribbin & John Gribbin
Brave the Wild River: The Untold Story of Two Women Who Mapped the Botany of the Grand Canyon, by Melissa L. Sevigny
Darwin's Most Wonderful Plants, by Ken Thompson
Abundant Beauty: The Adventurous Travels of Marianne North, Botanical Artist, by Marianne North
The Invention of Nature: Alexander von Humboldt's New World, by Andrea Wulf
Voyages of Discovery: A Visual Celebration of Ten of the Greatest Natural History Expeditions, by Tony Rice
The Essential Naturalist: Timeless Readings in Natural History, edited by Michael H. Graham, Joan Parker, and Paul K. Dayton
The Voyage of the Beagle, by Charles Darwin

Memoirs by Women in Plant Science

Drawing Botany Home: A Rooted Life, by Lyn Baldwin

Braiding Sweetgrass: Indigenous Wisdom, Scientific Knowledge, and the Teachings of Plants, by Robin Wall Kimmerer

The Plant Hunter: A Scientist's Quest for Nature's Next Medicines, by Cassandra Leah Quave

Lab Girl, by Hope Jahren

The Arbornaut: A Life Discovering the Eighth Continent in the Trees Above Us, by Meg Lowman

In Search of the Canary Tree, by Lauren E. Oakes

Finding the Mother Tree: Discovering the Wisdom of the Forest, by Suzanne Simard

To Speak for the Trees: My Life's Journey from Ancient Celtic Wisdom to a Healing Vision of the Forest, by Diana Beresford-Kroeger

ENDNOTES

PROLOGUE

1 Dwyer, J.D. (1957) *Androcalymma*, a new genus of the Tribe Cassieae (Caesalpiniaceae). *Annals of the Missouri Botanical Garden 44*(4): 295–297.

2 Christenhusz, M.J.M. & Byng, J.W. (2016) The number of known plant species in the world and its annual increase. *Phytotaxa 261*(3): 201–217.

3 Mora, C. et al. (2011) How many species are there on Earth and in the ocean? *PLoS Biology 9*(8): e1001127.

4 https://www.kew.org/

5 https://earth.stanford.edu/news/what-caused-earths-biggest-mass-extinction

6 Sidor, C.A. et al. (2013) Provincialization of terrestrial faunas following the end-Permian mass extinction. *PNAS 110*(20): 8129–8133.

7 McElwain, J. C., & Punyasena, S. W. (2007) Mass extinction events and the plant fossil record. *Trends in Ecology & Evolution 22*(10), 548–557. https://doi.org/10.1016/j.tree.2007.09.003

8 https://www.un.org/sustainabledevelopment/blog/2019/05/nature-decline-unprecedented-report/

9 Boehm MMA, Cronk QCB. (2021) Dark extinction: The problem of unknown historical extinctions. *Biology Letters 17*: 20210007. https://doi.org/10.1098/rsbl.2021.0007

10 https://www.theatlantic.com/ideas/archive/2020/08/women-scientists-have-evidence-about-sexism-science/615823/

11 Powell, K. (2021) The parenting penalties faced by scientist mothers. *Nature 595*: 611–613.

12 Sheltzer, J.M. & Smith, J.C. (2014) Elite male faculty in the life sciences employ fewer women. *PNAS 111*(28): 10107–10112.

13 Mora, C. et al. (2011) How many species are there on Earth and in the ocean? *PLoS Biology 9*(8): e1001127.

14 https://www.iucnredlist.org/assessment/supporting-information

CHAPTER ONE

1 Woolston, C. (2017) Graduate survey: A love–hurt relationship. *Nature 550*: 549–552. https://doi.org/10.1038/nj7677-549a

2 Bruneau, A., Mercure, M., Lewis, G.P., & Herendeen, P.S. (2008) Phylogenetic patterns and diversification in the caesalpinioid legumes. *Botany 86*(7): 697–718. https://doi.org/10.1139/B08-058

3 Prenner, G., and B. B. Klitgaard. (2008) Towards unlocking the deep nodes of Leguminosae: Floral development and morphology of the enigmatic *Duparquetia orchidacea* (Leguminosae, Caesalpinioideae). *American Journal of Botany 95*: 1349–1365.

4 Desmond, Ray (2007) *The history of the Royal Botanic Gardens Kew* (2nd ed.). Kew Publishing.

5 https://www.kew.org/science/collections-and-resources/collections/herbarium

6 https://www.nybg.org/blogs/science-talk/2019/11/pressed-for-time-herbaria-environmental-threats-assessment/

7 Lammertsma, E.I. et al. (2011) Global CO2 rise leads to reduced maximum stomatal conductance in Florida vegetation. *PNAS 108*(10): 4035–4040. https://doi.org/10.1073/pnas.1100371108

8 Endersby, J. (2008) *Imperial Nature: Joseph Hooker and the practices of Victorian science*. University of Chicago Press.

9 Thiers, B.M. (2020) *Herbarium: The quest to preserve and classify the world's plants*. Timber Press Inc.

10 http://sweetgum.nybg.org/science/docs/The_Worlds_Herbaria_2018.pdf

11 http://sweetgum.nybg.org/science/wp-content/uploads/2021/01/The_World_Herbaria_2020_7_Jan_2021.pdf

12 https://www.kew.org/science/collections-and-resources/collections/herbarium

13 Conversation with Ruth Clark, taxonomist at Kew Herbarium.

14 https://www.kew.org/kew-gardens/plants/living-collection

15 https://www.kew.org/read-and-watch/oldest-pot-plant-in-world-eastern-cape-giant-cycad

16 Huxley, R. (Ed.) (2019) *The great naturalists*. Thames & Hudson.

17 Endersby, J. (2008) *Imperial nature: Joseph Hooker and the practices of Victorian science*. University of Chicago Press.

18 https://naturalhistorymuseum.blog/2020/02/05/joseph-banks-the-banksian-herbarium-and-the-natural-history-museum-botany-collections/

19 https://www.fs.usda.gov/wildflowers/Rare_Plants/conservation/index.shtml

CHAPTER TWO

1 De Kock, C., et al. (2018) The functional role of the keel crest in *Polygala myrtifolia* (Polygalaceae) and its effects on pollinator visitation success. *South African Journal of Botany 118*: 105–111. https://doi.org/10.1016/j.sajb.2018.06.011

2 Prenner, G. & Klitgaard, B.B. (2008) Towards unlocking the deep nodes of Leguminosae: Floral development and morphology of the enigmatic Duparquetia orchidacea (Leguminosae, Caesalpinioideae). *American Journal of Botany 95*: 1349–1365. https://doi.org/10.3732/ajb.0800199

3 Bennett, J.A. (1989) The social history of the microscope. *Journal of Microscopy 155*: 267–280. https://doi.org/10.1111/j.1365-2818.1989.tb02890.x

4 Lightman, B. (2010) The microscopic world. *Victorian Review 36*(2): 46–49. doi:10.1353/vcr.2010.0006.

5 From Klosterman, C. (2022) *The nineties: A book* (ch. 1).

6 Taylor, I. (2011) University of British Columbia, Department of Botany Newsletter.

7 Kramer, A.T., Zorn-Arnold, B., & Havens, K. (2010) Assessing botanical capacity to address grand challenges in the United States (64 pp. plus appendices). https://www.bgci.org/wp/wp-content/uploads/2019/06/US-Botanical-Capacity-Report.pdf

8 https://www.etymonline.com/search?q=buxom&utm_campaign=sd&utm_medium=serp&utm_source=ds_search

9 Dwyer, J.D. (1957) *Androcalymma*, a new genus of the Tribe Cassieae (Caesalpiniaceae). *Annals of the Missouri Botanical Garden 44*(4): 295–297.

CHAPTER THREE

1 Cyranoski, D., Gilbert, N., Ledford, H. et al. (2011) Education: The PhD factory. *Nature 472*: 276–279. https://doi.org/10.1038/472276a

2 According to the National Science Foundation's most recently published data (2021): https://ncses.nsf.gov/pubs/nsf23300/report/u-s-doctorate-awards#overall-trends

3 Schillebeeckx, M., Maricque, B. & Lewis, C. (2013) The missing piece to changing the university culture. *Nature Biotechnology 31*: 938–941. https://doi.org/10.1038/nbt.2706

4 https://figshare.com/articles/dataset/2019_Nature_PhD_Students_Survey_Data/10266299

5 Cyranoski, D., Gilbert, N., Ledford, H. et al. (2011) Education: The PhD factory. *Nature 472*: 276–279. https://doi.org/10.1038/472276a https://www.science.org/content/article/first-us-private-sector-employs-nearly-many-phds-schools-do

6 Fernandes, J.D. et al. (2020) Research culture: A survey-based analysis of the academic job market. *eLife 9*: e54097 https://doi.org/10.7554/eLife.54097

7 Fernandes, J.D. et al. (2020) Research culture: A survey-based analysis of the academic job market. *eLife 9*: e54097 https://doi.org/10.7554/eLife.54097

8 Brischoux, F. & Angelier, F. (2015) Academia's never-ending selection for productivity. *Scientometrics 103*: 333–336. https://doi.org/10.1007/s11192-015-1534-5

9 https://figshare.com/articles/dataset/2019_Nature_PhD_Students_Survey_Data/10266299

10 https://www.science.org/content/article/first-us-private-sector-employs-nearly-many-phds-schools-do

11 https://figshare.com/articles/dataset/2019_Nature_PhD_Students_Survey_Data/10266299?file=18543281

12 Sousa-Baena, M.S., Sinha, N.R., Hernandes-Lopes, J. and Lohmann, L.G. (2018) Convergent evolution and the diverse ontogenetic origins of tendrils in angiosperms. *Frontiers in Plant Science 9*: 403. doi: 10.3389/fpls.2018.00403

13 Gerrath, J.M. & Posluszny, U. (2007) Shoot architecture in the Vitaceae. *Canadian Journal of Botany 85*(8): 691–700. https://doi.org/10.1139/B07-010

14 Barrows, C.W., Murphy-Mariscal, M.L., & Hernandez, R.R. (2016) At a crossroads: The nature of natural history in the twenty-first century.

BioScience 66(7): 592–599, https://doi.org/10.1093/biosci/biw043
15 Farber, P.L. (2000) *Finding order in nature: The naturalist tradition from Linnaeus to E.O. Wilson.* Johns Hopkins University Press.
16 Huxley, R. (Ed.) (2019) *The great naturalists.* Thames & Hudson.
17 Farber, P.L. (2000) *Finding order in nature: The naturalist tradition from Linnaeus to E.O. Wilson.* Johns Hopkins University Press.
18 Maslow, J. (1996) *Footsteps in the jungle* (ch. 1). Ivan R. Dee Inc.
19 Maslow, J. (1996) *Footsteps in the jungle* (ch. 4). Ivan R. Dee Inc.

CHAPTER FOUR

1 https://naturalhistory.si.edu/sites/default/files/media/file/2004-bdg-report-webpage2019edit.pdf
2 Darwin, C. (1989) *Voyage of the Beagle.* Penguin Books Ltd. (Original work published 1839)
3 Wulf, A. (2015) *The invention of nature* (ch.5). Vintage Books.
4 Gribbin, J. & Gribbin, M. (2008) *Flower hunters: Adventurous botanists and the lasting impact of their discoveries.* Oxford University Press.
5 Strimple, P. (1993) The green anaconda *Eunectes murinus* (Linnaeus). *Litteratura Serpentium 13*(2): 46–50.
6 Graduate thesis: Reyes, D., & Dever, J. (2007) Green anaconda *Eunectes murinus.* https://citeseerx.ist.psu.edu/document?repid=rep1&type=pdf&doi=f3c79a06378a5ca57ed49e0aa52f035b95bc66ab
7 Rivas, J.A. (1998) Predatory attacks of green anacondas (*Eunectes murinus*) on adult human beings. *Herpetological Natural History 6*(2): 157–159.
8 Mori, S.A. (1995) Exploring for plant diversity in the canopy of a French Guianan Forest. *Selbyana 16*(1): 94–98.
9 Britz, R., Hundsdörfer, A. & Fritz, U. (2020) Funding, training, permits–the three big challenges of taxonomy. *Megataxa 1*(1): 49–52.
10 Bebber, D.P. et al. (2010) Herbaria are a major frontier for species discovery. *PNAS 107*(51): 22169–22171

CHAPTER FIVE

1 Rice, T. (2017) *Voyages of discovery: A visual celebration of ten of the greatest natural history expeditions.* London Natural History Museum, London.

2 Rice, T. (2017) *Voyages of discovery: A visual celebration of ten of the greatest natural history expeditions.* London Natural History Museum, London.
3 Tobin, B.F. (1996) Imperial designs: Botanical illustration and the British. *Studies in Eighteenth-Century Culture 25*: 265–292.
4 Stagg, B.C. & Verde, M.F. (2019) A comparison of descriptive writing and drawing of plants for the development of adult novices' botanical knowledge. *Journal of Biological Education 53*(1): 63–78, doi: 10.1080/00219266.2017.1420683
5 Pardoe, H. & Lazarus, M. (2018) Images of botany: Celebrating the contribution of women to the history of botanical illustration. *Collections 14*(4): 545–566.
6 https://www.rcpe.ac.uk/heritage/elizabeth-blackwell-prison-plotting-and-curious-herbal
7 North, M. (2010) *Abundant beauty: The adventurous travels of Marianne North, botanical artist.* Greystone Books, Vancouver (reprint).
8 North, M. (2010) *Abundant beauty: The adventurous travels of Marianne North, botanical artist.* Greystone Books, Vancouver (reprint).
9 https://www.kew.org/read-and-watch/marianne-north-botanical-artist
10 Though it's worth noting that there is some disagreement among scientists as to the importance of key innovations in the overall scope of evolution.
11 Soltis P.S., Folk R.A., & Soltis D.E. (2019) Darwin review: Angiosperm phylogeny and evolutionary radiations. *Proceedings of the Royal Society* B 286: 20190099. http://dx.doi.org/10.1098/rspb.2019.0099
12 https://arboretum.harvard.edu/wp-content/uploads/2020/08/2013-70-4-beyond-pine-cones-an-introduction-to-gymnosperms.pdf
13 The name indicates an enclosed seed, as opposed to gymnosperms, which are named for their "naked" seeds.
14 Simpson, M.G. (Ed.) (2019) *Plant systematics* (3rd ed., pp. 167–185). Academic Press. https://doi.org/10.1016/B978-0-12-812628-8.50006-7
15 Soltis P.S., Folk R.A., & Soltis D.E. (2019) Darwin review: Angiosperm phylogeny and evolutionary radiations. *Proceedings of the Royal Society B 286*: 20190099. http://dx.doi.org/10.1098/rspb.2019.0099
16 Friedman, W.E. (2009) The meaning of Darwin's abominable mystery. *American Journal of Botany 96*(1): 5–21. 2009.
17 Silvestro, D., Bacon, C.D., Ding, W. et al. (2021) Fossil data support a pre-Cretaceous origin of flowering plants. *Nature Ecology and Evolution 5*:

449–457. https://doi.org/10.1038/s41559-020-01387-8

18 https://www.nybg.org/blogs/plant-talk/2012/03/exhibit-news/darwins-garden/darwins-star-orchid/

19 Breitkopf, H., Onstein, R.E., Cafasso, D., Schlüter, P.M. & Cozzolino, S. (2015) Multiple shifts to different pollinators fuelled rapid diversification in sexually deceptive *Ophrys* orchids. *New Phytologist 207*: 377–389. https://doi.org/10.1111/nph.13219

20 Fenster, C.B. & Martén-Rodríguez, S. (2007) Reproductive assurance and the evolution of pollination specialization. *International Journal of Plant Sciences 168*(2): 215–228

CHAPTER SIX

1 Steyaert, R.L. (1953) Etude sur les rapports entre les genres *Uittienia, Dansera*, et *Dialium* (Légum.-Caesalp.). *Reinwardtia 2*(2): 351–355.

2 Zimmerman, E. et al. (2017) Floral evolution and phylogeny of the Dialioideae, a diverse subfamily of tropical legumes. *American Journal of Botany 104*(7): 1019–1041.

3 Watson, L., and Dallwitz, M.J. (1993 onwards) *The genera of Leguminosae-Caesalpinioideae and Swartzieae: Descriptions, illustrations, identification, and information retrieval.* (Version 4, Aug. 2019) www.delta-intkey.com

4 https://today.duke.edu/2012/10/gagafern

5 Michiel Bussink, M. (2021) *Botanicus in oorlogstijd. Leven en werk van Hendrik Uittien*. KNNV. ISBN 978-90-5011-831-6; The author graciously fact-checked my passage on Uittien because I was unable to read his book, which is in Dutch.

6 van Steenis, C.G.G.J. (1948) *Dansera* and *Uittienia.* new Malaysian genera of the Caesalp.-Cassieae allied to *Dialium. Bulletin of the Botanical Garden, Buitenzorg, Ser. 3, 17*: 413–419.

7 The two are heavily interdependent, and though they are distinct, the terms are sometimes used somewhat interchangeably.

8 De Pasquale, A. (1984). Pharmacognosy: The oldest modern science. *Journal of Ethnopharmacology 11*(1): 1–16. doi:10.1016/0378-8741(84)90092-8

9 Pavord, A. (2005) *The naming of names: The search for order in the world of plants*. Bloomsbury USA.

10 Huxley, R. (Ed.) (2019) *The great naturalists*. Thames & Hudson.

11 Besse, P. (Ed.) (2014) *Molecular plant taxonomy: Methods and protocols* (ch. 1). Humana Press.

12 Browne, J. (1989) Botany for gentlemen. *Isis 80*: 593–621; This paragraph and the next largely from this paper. Browne is a noted Charles Darwin scholar.

13 Bewell, Alan. (2009) Erasmus Darwin's cosmopolitan nature. *ELH 76*(1): 19–48. Johns Hopkins University Press.

14 As quoted in: Shteir, A.B. (1996) *Cultivating women, cultivating science: Flora's daughters and botany in England, 1760 to 1860.* Johns Hopkins University Press.

15 Darwin, C. (1859) *On the origin of species by means of natural selection, or, The preservation of favoured races in the struggle for life.* J. Murray. London.

16 Endersby, J. (2009) Lumpers and splitters: Darwin, Hooker, and the search for order. *Science 326*: 1496–1499.

17 Besse, P. (Ed.) (2014) *Molecular plant taxonomy: Methods and protocols* (ch. 1). Humana Press.

18 Endersby, J. (2009) Lumpers and splitters: Darwin, Hooker, and the search for order. *Science 326*: 1496–1499.

19 https://www.kew.org/

20 Christenhusz & Byng (2016) The number of known plant species in the world and its annual increase. *Phytotaxa 261*(3): 201–217.

21 Mora C., Tittensor D.P., Adl S., Simpson A.G.B., Worm B. (2011) How many species are there on Earth and in the ocean? *PLoS Biology 9*(8): e1001127.

22 Costello, M.J. et al. (2013) Can we name Earth's species before they go extinct? *Science 339*: 413–416.

23 Costello, M.J. et al. (2013) Can we name Earth's species before they go extinct? *Science 339*: 413–416.

24 Locey, K.J. & Lennon, J.T. (2016) Scaling laws predict global microbial diversity. *Proceedings of the National Academy of Sciences 113*(21): 5970–5975.

25 Carruthers, J. & Robin, L. (2010) Taxonomic imperialism in the battles for Acacia: Identity and science in South Africa and Australia. *Transactions of the Royal Society of South Africa 65*:1, 48–64.

26 Kull, C.A. & Rangan, H. (2012) Science, sentiment and territorial chauvinism in the acacia name change debate. In Haberle, S.G. and David, B. (Eds.), *Peopled landscapes. Archaeological and biogeographic approaches to landscapes.* ANU Press.

27 Kull, C.A. & Rangan, H. (2012) Science, sentiment and territorial chauvinism in the acacia name change debate. In Haberle, S.G. and David, B. (Eds.), *Peopled Landscapes. Archaeological and Biogeographic Approaches to Landscapes*. ANU Press.

28 Engel, M.S. et al. (2021) The taxonomic impediment: A shortage of taxonomists, not the lack of technical approaches. *Zoological Journal of the Linnaean Society 193*(2): 381–387.

29 Hamilton et al. (2021) The future for a prominent taxonomy. *Insect Systematics and Diversity 5*(1): 1–2.

CHAPTER SEVEN

1 Clover, E.U. & Jotter, L. (1941) Cacti of the Canyon of the Colorado River and Tributaries. *Bulletin of the Torrey Botanical Club 68*(6): 409–419.

2 Bebber et al. (2010) Herbaria are a major frontier for species discovery. *PNAS 107*(51): 22169–22171.

3 By Little's own reckoning.

4 Just over 2.5 million for the 2021 competition.

5 Rodman & Cody (2003) The taxonomic impediment overcome: NSF's Partnerships for Enhancing Expertise in Taxonomy (PEET) as a model. *Systematic Biology 52*(3): 428-435.

6 Moyroud, E. & Glover, B.J. (2017) The evolution of diverse floral morphologies. *Current Biology 27*(17): R941–R951.

CHAPTER EIGHT

1 https://www.darwinproject.ac.uk/letter/DCP-LETT-13607.xml

2 https://www.thecrimson.com/article/2005/2/18/full-transcript-president-summers-remarks-at/

3 https://www.theguardian.com/uk-news/2015/jun/10/nobel-scientist-tim-hunt-female-scientists-cause-trouble-for-men-in-labs

4 https://www.science.org/content/article/mammoth-molars-yield-oldest-dna-ever-sequenced

5 Goulden, M., Mason, M. A., & Frasch, K. (2011) Keeping women in the science pipeline. *Annals of the American Academy of Political and Social Science 638*: 141–162. doi: 10.1177/ 0002716211416925; Though this isn't, perhaps,

the best metaphor, because it strongly suggests a passive issue, rather than institutions and individuals actively putting obstacles in women's paths. Some have referred to it as a "glass obstacle course."

6 Goulden, M., Mason, M. A., & Frasch, K. (2011) Keeping women in the science pipeline. *Annals of the American Academy of Political and Social Science 638*: 141–162. doi: 10.1177/ 0002716211416925

7 Goulden, M., Mason, M. A., & Frasch, K. (2011) Keeping women in the science pipeline. *Annals of the American Academy of Political and Social Science 638*: 141–162. doi: 10.1177/ 0002716211416925

8 Powell, K. (2021) The parenting penalties faced by scientist mothers. *Nature 595*: 611–613.

9 Ysseldyk, R. et al. (2019) A leak in the academic pipeline: Identity and health among postdoctoral women. *Frontiers in Psychology 10*:1297. doi: 10.3389/ fpsyg.2019.01297

10 Powell, K. (2021) How mothers get penalized in the scientific race. *Nature 595*: 611–613.

11 This section largely from: Shteir, A.B. (1997) Gender and "modern" botany in Victorian England. *Osiris 12*: 29–38; and the book by the same author: Shteir, A.B. (1996) *Cultivating women, cultivating science: Flora's daughters and botany in England, 1760 to 1860.* Johns Hopkins University Press.

12 Waring, S. (2018) *A sketch of the life of Linnaeus: In a series of letters designed for young persons.* Forgotten Books (classic reprint).

13 Endersby, J. (2008) *Imperial nature: Joseph Hooker and the practices of Victorian science.* University of Chicago Press.

14 Shteir, A.B. (1997) Gender and "modern" botany in Victorian England. *Osiris 12*: 29–38; Anne Witz as quoted in Shteir's Osiris paper: Defeminizing strategies turn "not upon the exclusion, but the encirclement of women within a related but distinct sphere of competence in an occupational division of labour."

15 Shteir, A.B. (1997) Gender and "modern" botany in Victorian England. *Osiris 12*: 29–38.

16 Parker, J.E. (2001) Lydia Becker's "school for science": A challenge to domesticity. *Women's History Review 10*(4): 629–650. doi: 10.1080/09612020100200303

17 As quoted in: Shteir, A.B. (1996) *Cultivating women, cultivating science: Flora's daughters and botany in England, 1760 to 1860.* Johns Hopkins University Press.
18 Gianquitto, T. (2013) Botanical smuts and hermaphrodites: Lydia Becker, Darwin's botany, and education reform. *Isis 104*(2): 250–277.
19 Becker, L.E. (1869) On the study of science by women. *The Contemporary review, 1866–1900 10*:386-404.
20 Gianquitto, T. (2013) Botanical smuts and hermaphrodites: Lydia Becker, Darwin's botany, and education reform. *Isis 104*(2): 250–277.
21 This paragraph: Gianquitto, T. (2013) Botanical smuts and hermaphrodites: Lydia Becker, Darwin's botany, and education reform. *Isis 104*(2): 250–277.
22 Gianquitto, T. (2013) Botanical smuts and hermaphrodites: Lydia Becker, Darwin's botany, and education reform. *Isis 104*(2): 250–277.
23 Huang, J., Gates, A.J., & Sinatra, R. (2020) Historical comparison of gender inequality in scientific careers across countries and disciplines. *PNAS 117*(9): 4609–4616; Bloodhart, B. et al. (2020) Outperforming yet undervalued: Undergraduate women in STEM. *PLoS ONE 15*(6): e0234685.
24 Cech, E.A. & Blair-Loy, M. (2019) The changing career trajectories of new parents in STEM. *PNAS 116*(10): 4182–4187.

CHAPTER NINE

1 Nichols, S. (2006) Why was Humboldt forgotten in the United States? *The Geographical Review, 96*(3): 399–415.
2 From Lightman interview.
3 From Lightman interview.
4 From Lightman interview.
5 From Lightman interview.
6 The entire book is available in the public domain here: https://www.gutenberg.org/ebooks/5726
7 Lightman, B. (2000) The story of nature: Victorian popularizers and scientific narrative. *Victorian Review 25*(2): 1–29.
8 Browne, J. (2002) *Charles Darwin: The power of place*. Princeton University Press.

9 Gianquitto, T. (2013) Botanical smuts and hermaphrodites: Lydia Becker, Darwin's botany, and education reform. *Isis 104*(2): 250–277.

CHAPTER TEN

1 https://wedigbio.org
2 https://digivol.ala.org.au
3 Vohland, K. et al. (Eds.) (2021) *The science of citizen science*. Springer Nature Switzerland.
4 Wulf, A. (2015) *The invention of nature* (ch. 3). Vintage Books.
5 Unlike Humboldt, Wallace wasn't independently wealthy. He made his way in the Amazon by selling some of the specimens he collected.
6 Aono, Y. and Kazui, K. (2008) Phenological data series of cherry tree flowering in Kyoto, Japan, and its application to reconstruction of springtime temperatures since the 9th century. *International Journal of Climatology 28*: 905–914.
7 The terms "community science" and "participatory science" are sometimes favoured due to the sensitive nature of the word "citizen," but since they are sometimes also used in other contexts, and because there hasn't been a consensus on this shift, I've used the term "citizen science" here to avoid ambiguity. I mean the term to be inclusive beyond the notion of national citizenship.
8 Maisonneuve, N. et al. (2009). NoiseTube: Measuring and mapping noise pollution with mobile phones. In: Athanasiadis, I.N., et al. (Eds) *Information technologies in environmental engineering. Environmental Science and Engineering*. Springer, Berlin, Heidelberg. https://doi.org/10.1007/978-3-540-88351-7_16
9 https://waspwatchers.umn.edu/about-wasp-watchers-program
10 https://phylo.cs.mcgill.ca
11 https://citizensciencegames.com/games/foldit/
12 Vohland, K. et al. (Eds.) (2021) *The science of citizen science*. Springer Nature Switzerland.
13 https://www.gbif.org/what-is-gbif
14 Irwin, A. (2018) No PhDs needed: How citizen science is transforming research. *Nature 562*, 480–482.

15 Di Cecco et al. (2021) Observing the observers: How participants contribute data to iNaturalist and implications for biodiversity science. *BioScience 71*: 1179–1188.
16 Personal communication, Peter Desmet, open data coordinator of LifeWatch team at the Research Institute for Nature and Forest.
17 MacPhail, V.J. & Colla, S.R. (2020) Power of the people: A review of citizen science programs for conservation. *Biological Conservation 249*: 108739.
18 https://www.zooniverse.org/about/publications
19 MacPhail, V.J. & Colla, S.R. (2020) Power of the people: A review of citizen science programs for conservation. *Biological Conservation 249*: 108739.
20 Theobald, E.J. et al. (2015) Global change and local solutions: Tapping the unrealized potential of citizen science for biodiversity research. *Biological Conservation 181*: 236–244. https://doi.org/10.1016/j.biocon.2014.10.021
21 Maund, P.R. et al. (2020) What motivates the masses: Understanding why people contribute to conservation citizen science projects. *Biological Conservation 246*: https://doi.org/10.1016/j.biocon.2020.108587
22 MacPhail, V.J. & Colla, S.R. (2020) Power of the people: A review of citizen science programs for conservation. *Biological Conservation 249*: 108739.
23 Maund, P.R. et al. (2020) What motivates the masses: Understanding why people contribute to conservation citizen science projects. *Biological Conservation 246*: https://doi.org/10.1016/j.biocon.2020.108587
24 https://citizensciencegames.com/games/
25 https://citizensciencegames.com/citizen-science-puzzle-game-in-bordelands-3/
26 Ceccaroni, L, et al. (2019) Opportunities and risks for citizen science in the age of artificial intelligence. *Citizen Science: Theory and Practice 4*(1): 29.
27 Campbell, C.J. et al. (2023) Identifying the identifiers: How iNaturalist facilitates collaborative, research-relevant data generation and why it matters for biodiversity science. *BioScience 73*: 533–541.
28 Campbell, C.J. et al. (2023) Identifying the identifiers: how iNaturalist facilitates collaborative, research-relevant data generation and why it matters for biodiversity science. *BioScience 73*: 533–541.
29 Callaghan, C.T. et al. (2021) Three frontiers for the future of biodiversity research using citizen science data. *BioScience 71*(1): 55–63.
30 MacPhail, V.J. & Colla, S.R. (2020) Power of the people: A review of citizen

science programs for conservation. *Biological Conservation 249*: 108739.

31 McKinley et al. (2017) Citizen science can improve conservation science, natural resource management, and environmental protection. *Biological Conservation 208*: 15–28.

32 This paragraph: Lughadha et al. (2020) Extinction risk to plants and fungi. *Plants People Planet 2*: 389–408.

33 https://www.iucnredlist.org

34 Lughadha et al. (2020) Extinction risk to plants and fungi. *Plants People Planet 2*: 389–408.

INDEX